2.50   Bishop Burto   639   D1761357

Carp and Pond Fish Culture

Carp and Pond Fish Culture

Second Edition

Including Chinese Herbivorous Species, Pike, Tench, Zander, Wels Catfish, Goldfish, African Catfish and Sterlet

László Horváth, MSc, PhD
Head of the Fishery Department, Szent István University, and Doctor of Science, Hungarian Academy of Science

Gizella Tamás (Horváth), PhD
Carp farm director, Fishery Scientist and author

Chris Seagrave, BSc (Hons)
Senior Lecturer in Fish Farming and Fishery Management, Sparsholt College, Hampshire, UK

Fishing News Books
An imprint of Blackwell Science

b

Blackwell
Science

Fishing News Books
A division of Blackwell Science Ltd
Editorial Offices:
Osney Mead, Oxford OX2 0EL
25 John Street, London WC1N 2BS
23 Ainslie Place, Edinburgh EH3 6AJ
350 Main Street, Malden
    MA 02148 5018, USA
54 University Street, Carlton
    Victoria 3053, Australia
10, rue Casimir Delavigne
    75006 Paris, France

Other Editorial Offices:

Blackwell Wissenschafts-Verlag GmbH
Kurfürstendamm 57
10707 Berlin, Germany

Blackwell Science KK
MG Kodenmacho Building
7–10 Kodenmacho Nihombashi
Chuo-ku, Tokyo 104, Japan

Iowa State University Press
A Blackwell Science Company
2121 S. State Avenue
Ames, Iowa 50014-8300, USA

First Edition published 1992
Second Edition published 2002

Set in 11/13pt Palatino
by DP Photosetting, Aylesbury, Bucks
Printed and bound in Great Britain by
MPG Books Ltd, Bodmin, Cornwall

DISTRIBUTORS

Marston Book Services Ltd
PO Box 269
Abingdon
Oxon OX14 4YN
(*Orders:* Tel: 01865 206206
        Fax: 01865 721205
        Telex: 83355 MEDBOK G)

USA and Canada
    Iowa State University Press
    A Blackwell Science Company
    2121 S. State Avenue
    Ames, Iowa 50014-8300
    (*Orders:* Tel: 800-862-6657
            Fax: 515-292-3348
            Web: www.isupress.com
            email: orders@isupress.com)

Australia
    Blackwell Science Pty Ltd
    54 University Street
    Carlton, Victoria 3053
    (*Orders:* Tel: 03 9347 0300
            Fax: 03 9347 5001)

A catalogue record for this title is
available from the British Library

ISBN 0-85238-282-0

Library of Congress
Cataloging-in-Publication Data is
available

For further information on Fishing News
Books, visit our website:
http://www.blacksci.co.uk/fnb/

Contents

The Authors

László Horváth

As a postgraduate László Horváth studied at two world famous establishments in Hungary: at the Fish Culture Research Institute, Szarvas (with Professor Woynarovich) and at the Warmwater Fish Hatchery, Szazhalombatta, where he eventually became Director of Research. Whilst there he was also involved in research and development for the Hungarian state fish farming industry as well as acting as a consultant in many developing countries.

Following a post as professor at the Agriculture University, Gödöllö, he established the Fishery Department at the newly named Szent István University in 1991 as its Head. He is also a Doctor of Science at the Hungarian Academy of Science.

László Horváth has published over 100 scientific papers and written several books on fish culture.

Gizella Tamás

Gizella Tamás also studied for her PhD at Szarvas. Married to László Horváth, she also worked at the Warmwater Fish Hatchery, Szazhalombatta, where she became Director of Production and a recognised expert on the technology of cyprinid fry rearing and the mass cultivation of zooplankton. After undertaking a managerial role for the Hungarian Federation of Fish Producers, she is now the director of a large carp farm in Hungary and the joint author of several books.

Chris Seagrave

After studying applied biology at Brunel University, Chris Seagrave left to carry out research into fish diseases at the Ministry of Agriculture, Fisheries and Food, Weymouth.

In 1980 he left to take up a teaching post as Sparsholt College as Senior Lecturer in the Department of Fish Farming and Fishery Management. He is still there!

In 1985 he developed a carp and ornamental fish farming business, which is currently one of the largest producers of pond fish in the UK. Chris Seagrave is also the author of textbooks on fish farming and fishery management.

Preface

The Hungarian fish farming industry is respected throughout the world for its expertise and innovations, especially in the spawning technology of fish, and the name of László Horváth is well known to all concerned with the farming of carp.

During 1980 László Horváth and his wife, Gizella Tamás, published two books on the culture of the common carp and other important fish farmed in Eastern Europe. The information in these books was originally aimed at technicians within the large Hungarian fish farming industry. The books sold well, and were regarded by many as essential texts for both farmers and students of fish farming alike.

During a visit to Hungary in 1988 we agreed that the books should be translated and published in English, as a text on the principles and practices of pond culture at this level was not readily available in the English language. The two books have been amalgamated here into one text, though some additions have also been made to make the book more relevant to a worldwide readership.

The final chapter has been enlarged to include a section on goldfish to cater for the increasing demand for information on ornamental species, although it should be emphasised that the culture of the common carp acts as a model for the majority of pond species.

I hope that, with the inclusion of a large number of original drawings painstakingly produced by L. Horváth senior, the book retains much of its Hungarian character.

Chris Seagrave
Editor

Preface to the Second Edition

The first edition of this book was produced in 1992. During the following decade the carp and pond farming industry has continued to grow at a pace, particularly in the Far East. As a food fish the carp continues to be extremely important but perhaps less so in Europe, as the major political changes experienced in the former Eastern Bloc countries have caused a wide range of social and economic changes.

In the UK the sport fishing industry has developed beyond all recognition, with the carp as the angler's main quarry. Intensively managed, densely stocked waters are now widely available for the pleasure angler. For the specimen hunter, fish as large as 20 kg (45 lb) are not uncommon.

The variety of species grown in ponds also continues to increase, for both the table market and the increasingly important ornamental fish trade. Technologically few major changes have been witnessed with the exception of the new 'synthetic' hormones now in common use in hatcheries around the world.

The first edition of this book was well received by the industry. It is hoped that this edition enjoys a similar reception.

Acknowledgements

The authors would like to thank the following for help in the preparation of this book: László Horváth senior for the delightful illustrations, Ferenc Muller for the translation into English, and Ann Levey and her colleagues at Sparsholt for typing the script.

   In addition, the authors would particularly like to thank Ildy Horváth for providing a number of additional illustrations in the style of her grandfather's originals.

Cover photograph reproduced courtesy of Mick Rouse of the *Angling Times*.

Chapter 1
Introduction

1.1 Characteristics of carp farming

In most countries of Europe and Asia, and in some areas of Central America, the common carp (*Cyprinus carpio*) is the most important cultured fish. Less importance is attached to this species in the Americas or Australasia.

Carp is one of a few species of fish that can be considered as domesticated, but there is a considerable difference between the domesticated form and its wild relation (from which it was originally bred) with respect to reproduction capacity, growth, utilisation of feeds, etc. The wild form is covered with scales and grows slowly while the 'noble' ones (i.e. domesticated forms of both scaly and mirror varieties) utilise artificially fed cereals and natural food well, giving rapid growth. Yields per unit area vary greatly, depending on the environment and rearing methods used. Using 'extensive' methods, about 0.5 tonnes/ha can generally be achieved whilst with the application of the most advanced technologies 2–3 tonnes/ha may be produced even in temperate climates. In the tropics where the growing season is longer, yields can be even higher still. Carp are highly resistant to handling, during harvesting, grading, transportation, etc., and to changes in water temperature and oxygen levels.

The palatability of carp is high, and the fish still enjoys considerable demand in the marketplace in most east European countries, in near and far east Asia and also among immigrant communities in the UK, USA and other western countries. It is extremely advantageous that as a farmed animal 50–60 per cent of the feed requirements of carp can be satisfied with cereals, the other 40–50 per cent being made up from small animals living in the ponds (inferior crustaceans, larvae of insects, molluscs, etc.). The reproduction capacity is extremely high and, during one season, 0.5–1 million fry may be produced from one female.

1.2 Historical background of carp husbandry

To understand and appraise appropriate methods of fish propagation and fry rearing it is necessary first to appreciate the historical background to fish culture and describe the 'simpler' methods of propagating carp. Early fish husbandry started in two centres – the ancient Chinese and Roman empires. Deliberate breeding practices created the basis for the development of successful propagation methods.

The earliest information regarding carp propagation comes from the ancient Chinese. The spawning of common carp stocked in ponds was observed and described as early as 451 BC, but for the Chinese carp early technology involved the collecting of eggs and fry from natural waters and then stocking them in fish ponds to be reared up to market size.

The first full account of European methods was documented in Europe in the sixteenth century by Dubravius, who described the principles of fish propagation and husbandry in fish ponds. In the eighteenth century a new phase of fish culture was initiated by Jakobi, who first fertilised trout eggs artificially. This work was a scientific curiosity in its time but was soon forgotten. The method was discovered again by Remi and Gehin a century later, and applied to fish farms. In 1851 the first fish 'seed' production farm was established in France. Propagation of salmonids was relatively easy due to the biological features of the species (the eggs are not sticky and they are also highly resistant to handling because of their thick shells), hence trout farms were established in several parts of the world. Garlick in the USA, as well as Nikolskiy and Vranskiy in Russia, developed the principles of trout husbandry and the results and experiences achieved were indirectly influential in the development of carp husbandry technology. In the nineteenth century Dubich and co-workers from Silesia in Eastern Europe, as well as their successors, performed outstanding work in developing carp farming methods, enabling the safe rearing and thus economic farming of carp to be possible. Dubich also observed that the natural production potential of fish ponds is closely related to the soil conditions, and he described the basic factors that influence the growth of carp in fish ponds. His method of carp propagation was the only one used for several decades in regions where sophisticated hatchery propagation techniques had not been introduced, and even today it is still the main method of seed production across the globe.

Detailed studies of the propagation and reproductive physiology of fish started in the 1930s. Gerbilskly and co-workers from Russia studied the endocrinal processes which regulate reproduction in fish (basing their work on that of Ihering in Brazil) and then elaborated a practical method for inducing ovulation. The principle of the method involves the use of the hypophysis (pituitary gland) which is transplanted to receptor fish, where exogenous gonadotrophic hormones released from the donor gland will trigger the ovulation process. With this method fish can be propagated to a preplanned schedule. This method enables mass propagation and scheduled production of seed stocks. The method known as 'hypophysation' was originally worked out for the propagation of sturgeonids, but it soon became evident that it can be applied successfully in the propagation of other fish species as well.

The hypophysation technique was used somewhat late for the propagation of carp (during the early 1940s). As the egg shells of carp contain chemicals that enable them to stick to vegetation in water, this 'stickiness' prevented carp eggs from floating freely in incubation jars, and hence made incubation in the hatchery impracticable. Therefore the hypophysation technique was used only to promote natural spawning in ponds, which is not a reliable method of producing seed.

A prerequisite for propagation in the hatchery was therefore to eliminate the stickiness of carp eggs. Several techniques were worked out for this at the end of the 1950s but the simple and efficient Hungarian method became the most widespread. This technique, the salt-carbamide (urea) treatment of eggs developed by Woynarovich from Hungary, is now applied all over the world. In recent years the traditional carp-orientated fish farming in Europe has been supplemented. Chinese practices, using several fish species together which better utilise the natural biocoenosis of fish ponds than achieved in monoculture, result in higher fish yields.

These additional species are: the grass carp or amur (*Ctenopharyngodon idella*), the silver carp (*Hypophthalmichthys molitrix*), and the bighead carp (*Aristichthys nobilis*). These are called (not really justifiably) 'herbivorous' fish and have been cultured for many centuries in China with fairly high yields. The seed stock for these species was collected from natural waters as they cannot propagate in conventional fish ponds. A favourable spawning environment can only be found in special river conditions. In the early 1960s Russian scientists such as Aliev worked out the method of artificial propagation for these species (now known as the Russian method) using the hypophysation technique with carp pituitary. This has led to the introduction of these species into many areas of the world. Later, Chinese experts elaborated an effective method using HCG (human chorion gonadotrophin) for broodstock injection, and utilising circular concrete tanks for egg and larval incubation (known as the Chinese method).

As in the case of many bred fish species simple spawning methods did not prove sufficient to guarantee a high production of seed. More efficient hatchery techniques had to be elaborated, which have since been further developed. Hypophysation can now be applied to many fish species for which natural spawning conditions cannot easily be provided.

As shown above, a rapid development has been achieved in propagation methods during past years, especially in recent decades. The main driving force is the need to satisfy the increased demand for seed material by the ever-developing fish farming industry.

1.3   The development of cyprinid farming in Europe

Fish farming developed considerably in Europe during medieval times as the Christian culture was flourishing. Monks and clergy living in monasteries were the first to introduce aspects of organised fish farming into Europe. The tradition of fish culture took shape predominantly in Central Europe (what is now Germany and the former Czechoslovakia) and the first mutated mirror carps were developed by selective breeding. Fish culture quickly divided into two distinct branches in medieval Europe. Trout farming developed mainly in the hilly/mountainous regions whereas in the warmer lowlands 'pond fish' culture took shape, based on the principle of using species of carp which demonstrated fast growth. Besides carp several other native species were kept, such as crucian carp and tench, as well as some predatory fish such as wels or pike perch.

As time progressed the two branches of fish culture developed simultaneously. Trout farming developed quickly in Western European countries where high rainfall and a well-developed industrial base encouraged technological change. Cyprinid farming, however, remained in an 'extensive' state based on the provision of natural food as the sole diet for the fish. At the same time in the drier eastern areas of Europe, where a continental climate with a three to four months' breeding and growing season and huge plains are found, ponds were built and carp breeding developed intensively. It has only been during the last decade that a higher proportion of cyprinid production in Western European countries, based on the use of high quality protein feeds, has developed.

Carp produced by this method are much more expensive, considering the increased production costs, than carp produced according to the Central European model where the fish meet their protein requirements from planktonic organisms of the pond. In Central Europe this cheap protein source would then be complemented with 'fodder' (inexpensive cereals). Carp produced in this way are generally more fatty, especially when the protein production (natural food) of the pond is not sufficient and the starch-based complementary feeding becomes predominant. In Central Europe, especially in countries with little or no marine fishing industries, people are used to eating fish produced in natural fresh waters as well as fish farmed in artificially built ponds. This provides the only cheap form of fresh fish available to the population and has ensured that there is a continued demand for pond fish.

Geographical features also rather favour pond fish culture in Eastern Europe. There is an abundance of land for the building of large (20–50 ha) fish ponds, the water supply can be provided by large lowland rivers and the continental climate is predictable and warm enough to stimulate the faster growth of the carp. A growing season of 100–150 days can be provided using water temperatures of 15–25°C, which are necessary for rapid growth.

Since the 1960s, the farming of the Chinese 'herbivorous' species (i.e. grass, silver and bighead carp) has spread quickly in some European countries and the proportion of these other species may reach 40 per cent of total annual production. The structure of fish consumption changes only very slowly as all cultures are conservative about such change. Thus, although these herbivorous species are very popular in the Far East, demand is currently still small in Europe and this limits their production. Similar marketing problems can also be found for the common carp. In the countries where there is little tradition of carp consumption the market demands did not expand at the same rate as production. This occurred in several Western European countries where, with the exception of some immigrant communities, the population is used to a fish diet composed only of trout, salmon and particular marine species. Generally they would find it difficult to accept carp. In contrast, in Central and Eastern Europe, fish soup and fried carp at Christmas is a very important part of the tradition, equivalent to the roast turkey in England.

1.4   Current status

Growing human population coupled with an increasing demand for protein encourages agriculture and aquaculture to develop and apply technologies that utilise natural resources more efficiently. From this point of view carp pond fish culture is one of the most effective and will continue to be exploited. In Europe this technology has been improved and maintained at a high level. The real gains, however, could come in those developing countries where there is a lack of protein (Africa and South America). Here carp culture could be complemented with particular local species even if some technological changes may be necessary due to different environmental conditions. Asia is not listed here since some of its countries were the first to develop cyprinid farming and many countries still continue efficient production today.

Production figures compiled by the FAO (*Fisheries Department Review* c886) suggest that the weight of fish produced in the aquaculture industry has increased dramatically, virtually doubling every 5 years. Carp, salmonids and tilapias dominate these statistics, and between them account for some 82 per cent of finfish production across the globe. A comparison can be seen in the box.

Annual production figures for finfish by group (in million metric tonnes)	
Carp	10.2
Salmonids	0.9
Tilapias	0.5

The global increase in carp production can be seen in Fig. 1.1. This upward trend has been mainly due to a massive increase in production in China, which completely dominates the figures.

Carp production figures by country (in million metric tonnes)	
China	8.30
India	0.26
Indonesia	0.14
Russian Federation	0.03
Ukraine	0.03

The countries listed are some of the main producers; compared with these, production in Europe is declining. This is partly as a result of the dramatic political reforms experienced in the former Eastern Bloc countries over the last decade, causing a whole range of social and economic changes.

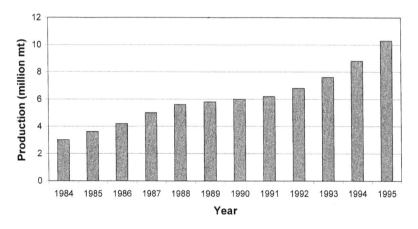

Global production of carp

Fig. 1.1

Angling potential

Finally, carp is valuable not only as human food but also as an angling fish. Even in countries where its flesh is not eaten, its value for angling is appreciated. This accounts for an important 'restocking' market for carp in certain countries, particularly the UK.

The last decade has seen an explosion in the number of privately owned fisheries in Britain. Most of these waters are highly stocked and well managed resources, catering for a range of angling practices. Such is their popularity that anglers are prepared to pay large sums of money to fish there. At the end of the day the catch is all returned back to the water alive.

Carp angling has thus become a very important recreational sport, satisfying the angler's demand for a large, hard-fighting fish.

Chapter 2
The Principles of Carp Farming

2.1 Ecological characteristics

Carp are thermophilic (warmth-loving), but will tolerate extreme, long-lasting cold as well as rapid fluctuations of temperature. The metabolism of carp and consequently its demand for food slows down gradually along with the decrease in temperatures, and practically stop at a water temperature of 4°C. The capacity for rapid growth which is characteristic of the species manifests best at a water temperature above 20°C. Carp shows high tolerance to variations in the ion concentrations of the water; it can live in brackish water as well as in alkaline waters of pH 9. It is also less sensitive to fluctuations in oxygen level, and can be efficiently cultured even at an oxygen concentration of 3–4 mg/l (fish kills may occur at oxygen concentrations as low as 0.3–0.5 mg/l). Carp can grow quite fast with occasional specimens reaching body weights of 20 kg. Carp feeds on benthic and zooplankton organisms, but also on seeds of plants and water weeds, detritus, etc.

2.2 Growth of carp in fish ponds

The biorhythms and growth of cultured fish species are not continuous in a temperate climate. Active (feeding) and passive (non-feeding) phases alternate in a regular sequence which in Europe is associated with changes in temperature.

Fish grow fast during their life stages of active feeding. This active period is during spring, summer and early autumn months when the water temperature remains steadily above 12–14°C. There is no growth during the non-feeding period; in fact, some loss in body weight may occur. During this period, fish withdraw to the bottom water strata at 4°C where, by maintaining their metabolism at its minimum level, they survive the cold winter season. Due to this seasonal change in activity, fish will only reach market size (1–1.5 kg) in a relatively long time (generally in three growing seasons).

Growth rates also change with age. The general rule in the realm of all living organisms, which is therefore valid for fish as well, is that the rate of growth is fastest up to sexual maturation, then it decreases until it completely stops. Fish species cultured in Central Europe tend to reach market size before sexual maturation and this rapid growth is characteristic of the whole production process. If, however, the growth rate is of the three growing seasons are compared, marked differences will be found. The fastest growth rate can be observed from the larval stage to the one-summer age. Growth here will represent a several hundred-fold increase over the initial weight. In the second growing season, ten-fold growth rates can be achieved under favourable conditions. In the third growing season, a five-fold weight gain is all that can be achieved. Nevertheless, this represents the highest growth in absolute terms, averaging 2 kg compared with the 200 g and 20 g weight gain achieved respectively in the second and first growing seasons (Table 2.1).

The growth rate naturally is not determined solely by temperature and/or age of fish. It is significantly influenced by a number of other factors, e.g. stocking density, quality and quantity of food, oxygen concentrations, etc.

Table 2.1 Seasonal growth of common carp and Chinese herbivorous carp in a temperate climate.

Production stage	Initial weight (g)	Final weight (g)	Growth from initial weight
From fry to advanced fry	0.003–0.005	0.3–1.5	100–300-fold
From advanced fry to one-summer fish	0.3–1.5	20–70	50–100-fold
From one- to two-summer fish	20–70	100–350	7–10-fold
From two-summer to market fish	100–350	600–1750	5–7-fold

2.3 The farming cycle

Fish farming can essentially be divided into two fundamental processes:

(1) the propagation and nursing of fry;
(2) the rearing of the growing fish.

There are pronounced differences between the two stages, both in principles and practice. The first stage requires a sound biological knowledge of the fish in terms of reproduction biology, ecology, taxonomy, etc., while in the second stage a knowledge of the technological, nutritional, physiological, and management practices is necessary.

The first stage, i.e. the propagation and rearing of fry, involves the preparation of quality broodstock which are spawned using natural techniques or sophisticated hatchery procedures. Feeding fry are stocked into specially prepared ponds. After 3 to 4 weeks fingerlings are harvested and transferred to ongrowing ponds later in the season. The rearing of one-summer-old and/or two-summer-old fish generally starts earlier with fish stocked out from wintering ponds. This operation is performed as the weather warms up after the winter, with both age groups being stocked into carefully prepared ponds. As the water warms up during late spring and early summer, fish start feeding and become active and the process of growth will recommence. During this period fish will convert both natural food from the pond and supplementary feed into fish flesh. As the year turns once again to autumn, along with the decrease of water temperature, the appetite of the fish will decrease and be at a minimum level by, say, the beginning of November. At this time the passive 'wintering' period will begin. Under subtropical/tropical conditions the practices of carp farming are different from those described previously. Water temperature in these regions is high enough to maintain the active feeding of carp for most of the year. Water cools down only for a short period of time if at all. Thus, fish keep their appetite and their growth will be continuous! If the growth rates of the common carp under these conditions are compared with those in warmer climates, it can be seen that carp will reach the same body weight (mass) in one year rather than the three years required in the temperate zone. There is also a difference in the weight of market-sized fish demanded under tropical conditions.

There are regions where a fish with a body weight of 300–500 g is large enough to market. This body weight is reached in a relatively short period of time, which results in a much shortened growing season.

2.4   Farm production

A fish farm as an economical unit consists of fish ponds in which fish husbandry is performed with scheduled and well planned activities. According to the fish yield of the farm one can distinguish between:

(a)   Extensive fish farming (Fig. 2.1), where stocking density is generally low; there is hardly any or even no supplementary feeding, and natural production results in some several hundred kilograms per hectare per year (see Table 2.2).

(b)   Semi-intensive fish farming, where the stocking is higher, management is more sophisticated and daily feed is supplemented. Yields can be as high as several thousand kilograms per hectare per year.

SPRING

AUTUMN

Fig. 2.1 In its simplest form 'extensive' culture involves the stocking of fingerlings into a pond. These are later harvested after the growing season.

Table 2.2 Natural productivity found in a range of farm ponds (kg/ha/yr).

Definition of pond	Carp monoculture (kg)	Cyprinid polyculture (kg)
Poor	100	150
Moderate or medium	100–200	150–375
Good	250–400	375–600
Excellent	400–600	600–900

Within the two above described forms of fish farming, stocking structure can be monocultural (one species, generally common carp) or polycultural (several non-competitive species, generally cyprinids). In both monocultural and polycultural stocking structures, the aim of the farmer is to choose fish species utilising the natural protein source of fish ponds. Since this source is rather limited, fish yield resulting from natural production of fish ponds can be increased with inorganic or organic fertilising.

Fish yield can also be increased considerably by supplementary feeding, though the quality and quantity of food used makes a significant difference. Carp is an omnivorous species, and will consume cereal grains of high carbohydrate content (Fig. 2.2). The development of the carp will be good if the ratio of natural and supplementary food is one to one. Natural food provides indispensable amino acids, fatty acids and various vitamins, while cereal grains – due to their high starch content – provide the energy for rapid weight gain.

Thus, the fish yield can be attributed principally to yield from the natural food production of fish ponds, plus yield from the supplementary food (cereal grain) (Fig. 2.3). Carp can also be cultured, like trout, under intensive conditions at a very high stocking density and fed on complete, pelleted feeds (Fig. 2.4). Such intensive systems are carried out in 'recycling systems' where fish are kept in tanks, and this also includes certain types of cage culture systems. However the cost of carp produced in these systems is much higher than those reared in ponds, since food of high starch content is not in itself enough for the growth of common carp. If cereals were used as the only source of food, deficiency diseases could easily develop. A complete diet for the common carp should therefore contain all the protein, fat and vitamin components provided by natural food. Thus, the production cost in super-intensive fish farming

Fig. 2.2 During early farming practices carp were produced in monoculture. The basis of production was the use of inexpensive cereal feeds.

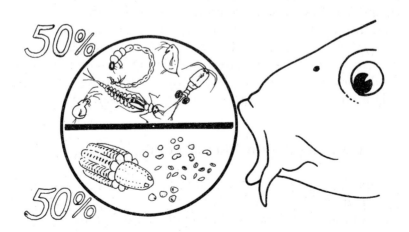

Fig. 2.3 Fish yields from ponds can be attributed to food derived from natural food sources plus that from the supplemented diet (cereals, etc.).

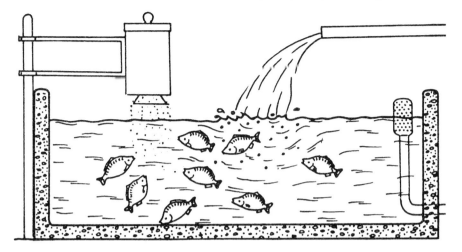

Fig. 2.4 Carp can be produced intensively using complete diets, but this is often an uneconomical way of producing fish meat.

systems is greatly increased by the cost of expensive complete diets, as well as the running costs of these sophisticated facilities (heating, water filtration, aeration, pumping, etc.).

Besides providing a habitat (medium) for fish, fish ponds are cheap protein-producing 'factories' where processes resulting in the production of food organisms for fish can be easily controlled. The common carp, due to its feeding habit, consumes only certain kinds of food organisms in the pond (primarily invertebrate animals). This therefore leaves large quantities of plant material unutilised. Other fish species that feed on phytoplankton and macrovegetation (reeds, submerged plants, etc.) can be additionally stocked in a polyculture system without competing with the common carp for food. For this purpose, the herbivorous fish species of Far East origin are suitable (i.e. grass, silver and bighead carp).

2.5   Planning the stocking of ponds

To optimise the production of a farm, the stocking and cropping plan will have to consider many factors as the individual growth rates of fish stocked into ponds are determined by:

- the number of fish stocked into a given area
- their size/weight at the time of stocking
- the *quality* of feed provided

The market requirements will also dictate the numbers and sizes of fish required.

(1)   The *higher the number* of fish stocked, the greater will be the competition for food and the *lower will be the expected individual weight* prior to harvest. This can be explained by the fact that every pond has a production 'capacity'. This capacity depends upon many factors including:

- the properties of the pond 'soil' layer;
- water quality;
- nutrient availability;
- pond depth;
- climate (i.e. temperature).

The production capacity of a pond is very difficult to determine and therefore it is impossible to provide a general formula or principle that can be applied to all ponds. Instead, farmers have to rely on past experiences and local conditions and to be able to recognise the production capacity of a pond based on expertise gained over many seasons.

(2)   The larger the fish is at the time of stocking, the larger it can grow over the summer period. If a farm does not have enough fish of a particular size to meet production targets, additional stock will have to be purchased from elsewhere to supplement supplies.

(3)   Increased individual weights of fish can be controlled by varying the nutrition of the feed. In 'extensive' culture the necessary requirements of the fish are met by 'natural' foods but as the pond culture becomes more intensive, supplementary feeds need to be introduced. When very high yields are required, 'complete' diets can be supplied in pelleted form, but inevitably feed costs will be significantly higher than when cheap cereal supplementary feeds are used. The profitability of this type of culture depends on the value of the stock at market and the price of the feeds. These factors vary considerably in individual countries.

In general it is possible, using 'semi-intensive' production technology in shallow ponds in a temperate climate, to sustain 1000–2000 kg of fish/ha safely. Using this figure, pond production for a '1 hectare' pond can be planned using the following example

Initial weight of two-year-old (C2) fish:	250 g
Final required weight of individual fish:	1000 g
Expected pond capacity:	1500 kg
Total number of fish required at stocking:	1500 kg ÷ 1000 g = 1500 fish
Allowing for 20% loss, total stocking:	1500 + 20% = **1800 fish**

2.6   The fish farm

Fish farms are structured according to the specialisation of particular units. The overall production unit is called the fish farm or, in Eastern Europe, a *pond fishery estate*. It can be an individual establishment or part of a larger agricultural development. The pond fishery estate may be made up of smaller units called *pond units*, the basic production unit being the *pond*. Estates are characterised by the farming practice which is undertaken, i.e. by species, age group, etc. A range of enterprises exists, from a 'full operation' to a 'part operation' farm. A 'full operation' farm produces all fish age-groups from fry, advanced fry, fingerlings to marketable size fish (Fig. 2.5), whereas a 'part operation' farm confines its operation to certain phases of the production cycle only.

The definable phases of the full operation production are as follows:

(1)   Hatchery propagation, which starts with the procurement of eggs and ends with 'feeding fry'.
(2)   Early fry rearing, which starts with the stocking of feeding fry and ends with one-month-old 'advanced fry'.
(3)   Fingerling rearing, which starts with the stocking of one-month-old fish and ends with the harvesting of fingerlings.
(4)   Ongrowing (fattening) period from the stocking of fingerlings until the end of the second season.
(5)   Production of marketable size fish, from the stocking of two-summer-old fish until the end of the third season (except in the case of a two-year growing cycle).

The first three stages, if outlined under *extensive* farming practices, can be done in ponds. They may still operate all of the above stages and are defined also as 'full operation'.

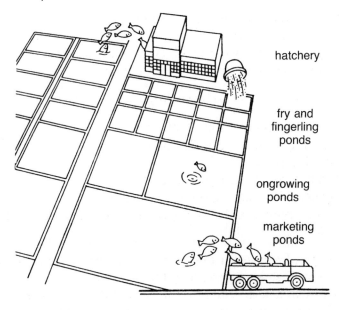

hatchery

fry and
fingerling
ponds

ongrowing
ponds

marketing
ponds

Fig. 2.5   On full operation farms, all stages of the process are found from propagation to final product (market size fish).

Specialised farms also exist that operate only some of the production stages. Such enterprises may be *fry rearing farms* or *fingerling producers* which operate the first three phases (propagation, advanced fry and fingerling production). Other farms may only ongrow fingerlings to marketable size, buying new stock annually. These two types would constitute the most common types of 'part operation' enterprises.

2.7   The ponds

Carp and other pond fish grow well in shallow ponds that can warm up quickly in the summer. In Europe and subtropical regions, ponds should be 1–1.5 metres deep, but in very hot climates they may be constructed deeper (2–2.5 m) to prevent the water from becoming too warm.

Pond sizes are chosen according to the aim of the production unit and the land available for development. Ponds can be found from fry ponds of a few hundred square metres to 100-hectare extensive production ponds. On a full operation farm each stage of production is represented by the following proportions of total farm area:

- One-summer fingerling production (known as C1 production): 10 per cent
- Ongrowing fingerling production (C2): 20 per cent
- Market fish production (C3): 70 per cent

Ponds are differentiated according to size:

- Fry rearing ponds: 100–5000 m²
- Fingerling ponds: 5000–20 000 m²
- Fattening ponds: 5–10 ha
- Market size production: >5 ha
- Specialised wintering ponds: 600–2000 m²
- Storage ponds: 0.2–10 ha (which may also be used for wintering)

Ponds are also used in the hatchery unit for keeping broodstock as well as spawning (e.g. Dubich ponds).

Ponds can also be classified in different ways according to how they are constructed. There are those formed by flooding valleys in hilly regions, and those formed in flat ground by building up banks. Ponds can be constructed in valleys by damming a stream between hills (Fig. 2.6), enabling more ponds to be constructed further up the valley. The water supply under these circumstances is good for pond culture as it usually contains high quantities of nutrients washed in from surrounding agricultural land. This promotes the development of large amounts of natural food.

The disadvantage is that the water supply may be unreliable and dependent on rainfall.

If the water flows from one pond to another, disease may readily spread. It is more effective to divert the stream at the side of the valley where excess water flows around and not through the pond (Fig. 2.7). Pond filling can then be controlled. The depth of water in such ponds is variable. It may be 2–3 metres at the deepest end and very shallow at the upper end.

Ponds built on flat land are built close to the water source (river, canal, reservoir), usually on land of poor quality (marshy) which is unsuitable for other agricultural purposes.

As the natural productivity of a pond is determined by the soil type, these ponds are generally less fertile and natural food production will be lower compared with ponds built in valleys. To construct the ponds, soil is banked up from the bottom to form the dams. The depth should be uniform (1–1.5

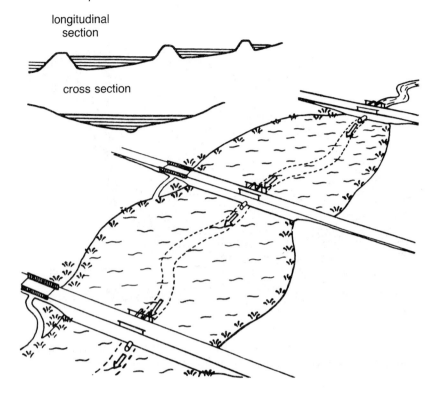

longitudinal
section

cross section

Fig. 2.6 Fish ponds created by damming valleys are characteristic of particular areas. Ponds are formed in sequence down the valley.

metres). There may be an external harvesting area which can be utilised by several adjacent ponds (Fig. 2.8).

Ponds do not have particular requirements for water quality. Most supplies are suitable for filling ponds as long as there is no evidence of pollution from upstream industrial effluents (Table 2.3). Water requirements are highest during the spring at the time of filling, so ponds constructed in valleys are usually filled during the autumn to ensure they can be totally filled.

Additional water is needed every year to replace that lost by seepage and evaporation but where farming is intensive additional water is also needed to provide throughflow and even to provide complete water changes (to maintain water quality within the pond). If the topography allows, inlet channels may provide water for ponds by gravity. If this is not the case, pumps are required; the cost of pumping water must be taken into consideration regarding the viability of the

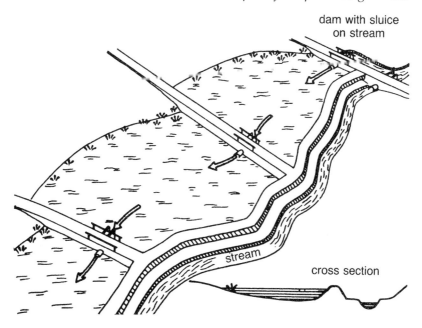

Fig. 2.7 Diversion ponds. It is preferable to run the inlet stream adjacent to the ponds rather than flowing through them.

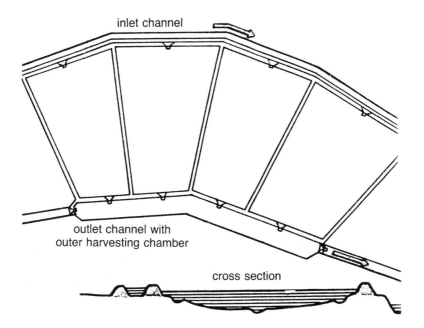

Fig. 2.8 Ponds constructed on flat plains can cover large areas but their production capacity may be lower.

Table 2.3   Water quality criteria.

	Unit	Optimal level (for hatcheries)	Permissible limit level (for pond farming)
Conductivity	μS	1000–2700	up to 6000
H_2S	mg/l	0.0	0.1
NO_2	mg/l	below 0.5	0.5
NO_3	mg/l	1.0–10.0	15.0
NH_4^+ ion	mg/l	1.5–2.0	3.0
Free NH_3	mg/l	0.0	0.1
Saturation of O_2	%	above 70	50
Dissolved O_2	mg/l	5–12	3–4
pH	–	7.0–8.5	9.0
Fe, Mn	mg/l	below 0.02	0.02

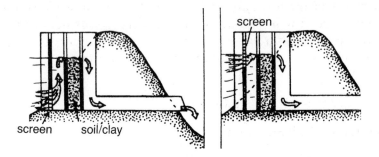

Fig. 2.9   A monk is a traditional drainage system for farm ponds. Water levels are controlled using wooden boards. Excess water is filtered out with a screen of suitable size.

farming operation. Drainage by gravity is an essential requirement of a good pond and is usually controlled by a monk (Fig. 2.9).

The maintenance of inlet and outlet channels is very important. If they are likely to become blocked with plants, two- or three-year-old grass carp may provide a good controlling agent at approximately 250 kg/ha.

In areas where seasonal water flow is variable, fish farms should not be constructed unless water flow is adequate for 80 per cent of the year. Sites with lower flows than this may not be viable for development.

Chapter 3
The Technology of Seed Production

Throughout the world a wide variety of methods is used to produce carp fry. Many are based on simple techniques unchanged over hundreds of years, whereas in contrast, others involve sophisticated technology used to produce vast numbers of fry throughout the year. The methods chosen to produce seed will depend upon the resources of the farm or country and also upon the economics of the enterprise. Each method has its own advantages and disadvantages.

3.1 Extensive methods

Spawning in large ponds

This is a widespread and simple method of propagation but is only applicable to newly constructed or previously dried ponds with an area of vegetation. These grass-covered areas will provide ideal spawning places after filling. Ponds which develop a growth of aquatic plants may also be suitable.

When the water starts warming up, ponds are filled and previously selected breeders can be stocked (two to three fish per hectare). Generally two to three males are stocked with each female. Under favourable conditions (i.e. suitable temperatures and the presence of a suitable spawning medium) good results can be achieved (Fig. 3.1). It is, however, very hard to check the amount of eggs laid and the fertilisation rates. It would be mid-summer before the success of spawning could be determined but by then spawning cannot be repeated.

This method of spawning should be avoided, if possible, since it will not provide the quantity of seed material necessary for large-scale fish farming. As a reserve production using simple facilities, or to provide cheap seed materials, however, it is suitable.

Fig. 3.1 In favourable conditions natural or wild spawning may result in inexpensive fry production, but the risks of failure are very high.

Spawning in small 'Dubich' ponds

The essence of the technique (previously outlined in Chapter 1) is that all the natural factors necessary to induce carp spawning are provided under pond conditions, i.e. rapidly warming shallow water (18–20°C) with macrovegetation on the bottom for spawning, sufficient dissolved oxygen, presence of both sexes, etc. (Fig. 3.2). Small shallow ponds are necessary for the Dubich technology, the best being 120–300 m² with a water depth of 30–60 cms, located in a protected area of the farm. These small ponds are kept dry when not in use, with a managed grass covering on the bottom all year round. When the water temperature reaches a steady 18–20°C, ponds can be prepared. They have to be cleaned, the grass cut, and then filled with filtered oxygen-rich water up to 25–30 cms in depth. Spawners are then stocked, two or three females and four or five males, and the water level raised slowly up to 50 cms. One or two days after stocking, spawning will occur (under adverse conditions spawning will not happen).

Parent fish are removed immediately after spawning to avoid any harm to eggs that have been laid. The water level

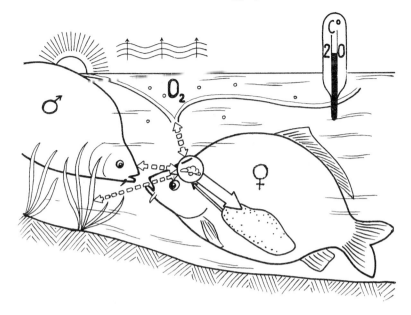

Fig. 3.2 The parameters required to create the correct spawning environment stimulate the release of hormones, which control spawning behaviour.

should be lowered rapidly, to drive fish to the deeper parts of ponds, from where they can be netted carefully. The water level is then rapidly raised to keep all the eggs under water. The eggs that have been laid are firmly attached to the grass and, depending on the temperature, will hatch in four to eight days. The number of hatched and pigmented larvae can be estimated using the so-called 'white plate' method (see Chapter 4). Ten to twelve days after hatching, when larvae are approximately 12–15 mm in size (so-called 'mosquito carp') they have to be harvested and stocked into other, bigger ponds for continued nursing (Fig. 3.3).

With this method, which is rather time-consuming and labour-intensive, seed material can be produced with good success. However, the results are greatly influenced by the weather. The amount of larvae produced by this method is much higher than under natural conditions, but still a lot of eggs and larvae are consumed by water micro-organisms (fungi, bacteria), carnivorous crustaceans (copepods), and water insects, birds, etc.

This method is still currently used in pond fish farms, not only in Europe but in some subtropical and tropical regions as

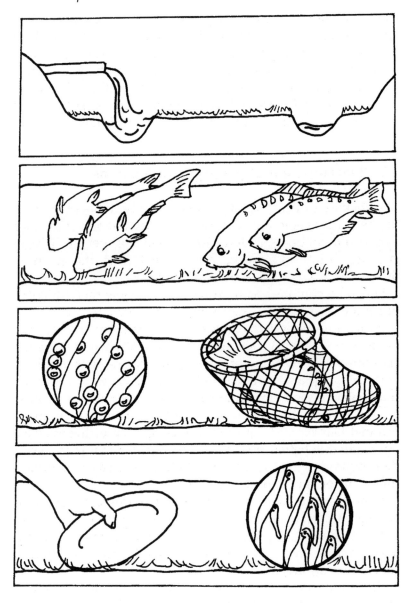

Fig. 3.3 The Dubich pond method of fry production involves a 'controlled natural method' which is not always reliable in a temperate climate.
(1)   The Dubich pond is flooded.
(2)   Broodstock are introduced.
(3)   After successful spawning broodstock are removed.
(4)   Fry hatch out in the pond.

well. Elaboration of the 'hypophysation technique' (described later) facilitated the further improvement of the Dubich method. If hypophysis-treated (i.e. hormone-treated) spawners are stocked into the well-prepared Dubich ponds, spawning almost always occurs. By this method, however, only spawning is ensured; the eggs and larvae still remain without protection, and thus the survival rate is not improved.

3.2   Development of propagation methods

The stickiness of carp eggs presented a substantial obstacle to further improvement of the spawning technology since carp eggs could not be incubated and hatched in Zuger jars. (Zuger jars are 7-litre incubation jars, named after the town in Switzerland, Zug, where they originated.)

Several techniques were tested to overcome this difficulty, but none of them met the requirements for large scale seed production. In one such method, spawners are stocked into Dubich ponds after hormonal treatment, then just before spawning, fish are taken out, stripped, and the eggs fertilised. Fertilised eggs are then attached to baskets, linen or any other material suitable for them to stick to and are hatched in tanks or small ponds. This method is an advance compared with the original Dubich method in that development of the attached eggs can be better observed, the fertilisation ratio easily determined and, with daily malachite green treatment, fungus infections of eggs can be avoided.

Methods applied in subtropical and tropical regions employ similar principles, where carp are spawned in cages of fine net-cloth (hapas) (Fig. 3.4) or in small ponds where eggs are laid on the fine fibres of water weeds or other fibrous materials (Fig. 3.5).

For a long time the Dubich method, whereby the natural spawning environment is imitated on the fish farm, was the most reliable way of producing carp fry. It is still used unchanged in some areas even now. However, in recent decades, increased knowledge of hydrobiology and nutritional biology has created possibilities for a considerable increase in carp stocking rates per unit area of pond. This has resulted in an increased demand for fry. To meet the increased demand, more intensive methods of propagation have been developed. As already stated, the original Dubich method did nothing to

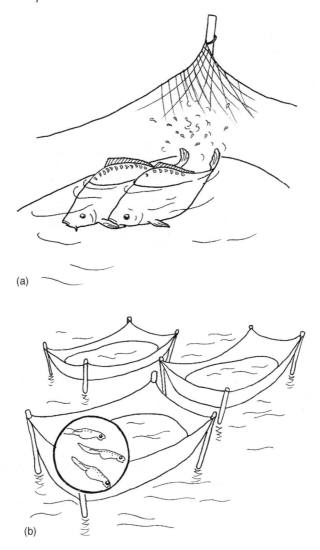

(a)

(b)

Fig. 3.4   (a) Broodstock are introduced into the 'hapa'. (b) After spawning, broodstock are removed and fry hatch inside the cage.

secure the survival of eggs and larvae, so the introduction of methods for protecting these delicate organisms was a necessary step forward. In addition, new techniques for controlling and inducing spawning were also developed, for example:

(a)   Spawners are captured from the natural environment during the spawning season (Fig. 3.6), ova and sperm are

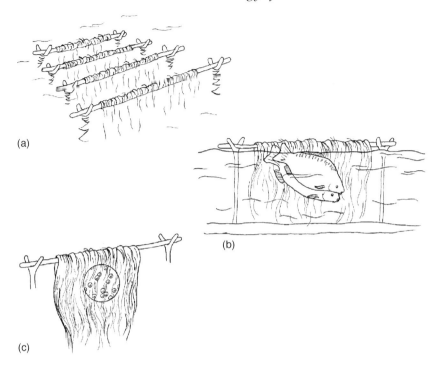

(a)

(b)

(c)

Fig. 3.5 (a) Spawning media in the form of mats or ropes are introduced into the broodstock pond. (b) After spawning, eggs are attached to the fibres of the medium. (c) The mat is then removed to a prepared pond where the dry hatch out.

stripped and fertilisation carried out by the 'dry method' (i.e. mixing sperm and eggs in a dry container before adding water or a fertilising solution to stimulate sperm motility and egg fertilisation). Eggs are incubated while attached to a suitable substrate. Using this technique, however, only partial protection of the delicate eggs and larvae is possible.

(b)   Investigations also began with the aim of perfecting a system for controlled propagation by inducing ripe broodfish to spawn by treating them with gonadotrophic hormones taken from other fish (Fig. 3.7). Initially, induced spawning following pituitary treatment was used only to ensure that spawning in ponds occurred at the required time. Spawners were treated before being transferred to ponds for natural spawning. This method served only to secure the safe production of fertilised eggs.

Fig. 3.6 Brood fish can be captured as they undergo spawning in the wild. Eggs are stripped out and fertilised immediately and then either transferred to the hatchery or introduced onto spawning media in prepared ponds.

Fig. 3.7 The spawning environment can be substituted by an injection of pituitary extract.

(c)   As a synthesis of the above two techniques, a method was elaborated by which ovulation was induced; fish were stripped and eggs fertilised by the dry method. This was a considerable advance, as it allowed the fertilised eggs and newly-hatched larvae to be kept in a controlled, protected environment.

For a long time, care of eggs in the hatchery was hampered by the fact that carp eggs become sticky when they come into contact with water, and consequently it proved impossible to incubate them in jars. Therefore, finding a way to eliminate egg stickiness was essential for the development of advanced techniques for propagating carp. Several methods were developed for this purpose, the simplest and most effective being the salt-carbamide method elaborated by Professor Woynarovich of Hungary and applied today in many countries (see section 3.5). In the following chapter, the most up-to-date propagation method is described. This method is widely used in fish farms and is the most effective system currently available.

3.3   Broodstock maintenance

Successful fish propagation by any technique can only be achieved by using suitably prepared broodstock. The reproduction process makes high demands on the body resources of the fish and propagation is possible only in broodstock that have been specially prepared. The necessary conditions for broodstock preparation must be maintained artificially during their preparation time. The fish breeder's task is therefore to create the conditions necessary for the development and maturation of the gametes.

Mature fish that have previously been successfully propagated (one or more times) require particular management during the annual cycle of egg development. The season begins immediately after spawning. During the summer female broodstock are kept in low density (100–300 fish/ha) in a 'natural' environment. To develop the eggs quickly within the ovary, females require a high input of nutrients which increases as the ovary develops. These would normally be provided by the natural organisms within the pond. Nutrient availability can be increased in two ways:

(a) by adding fertiliser to the pond to increase the produc-
     tion of food organisms;
(b) by feeding the females directly.

The availability of food organisms can be increased by using
artificial (inorganic) fertilisers, but zooplankton does not
satisfy the complete nutritional requirements of the fish when
they are feeding vigorously. The fish farmer should also
provide a supplementary feed in the form of cereals (wheat or
barley), at approximately 2 per cent of body weight daily.

After spawning, fish leave the hatchery in a weak,
exhausted condition. Even when considerable care is taken
some fish may be damaged or may die by becoming egg-
bound.

Surviving stock lose a significant percentage of body weight
(5–10 per cent) during the propagation process. Fish instinctively try to replace this loss and consequently their appetite is
vigorous in this early period. As nutrients from feedstuffs are
taken up very quickly into the newly developed eggs, energy-
rich feeds can be fed without causing excess development of
fat. Feeds such as grain or sprouted barley can be safely fed to
recently spawned fish. It must be noted, however, that this
food alone is not of good enough quality to produce eggs.
Good ovary development can only be achieved if natural food
is also abundant to provide broodstock with vitamins and
fatty acids.

It may also be advisable to stock some small carnivorous
fish with the broodstock carp, i.e. two-summer-old zander or
wels catfish, to eliminate any juveniles produced by wild
spawning (Fig. 3.8).

By the time autumn is reached sexually mature carp will be
ready to spawn. The ovary has completed development and
contains mature eggs – however, with the fall in water temperatures the metabolism of the fish gradually slows down as
the carp prepare for overwintering. Usually broodstock ponds
are harvested in the autumn. Fish are carefully selected for
breeding the following spring. It is a specialist skill to judge
the condition of the fish and determine whether the fish are in
a healthy state or require any disease treatment. During the
selection process, substandard fish are eliminated and mar-
keted for food. Suitable brood fish are stocked into a wintering
pond. For the wintering of broodstock, deep ponds with a
good supply of freshwater are used. Herbivorous fish require

Fig. 3.8 Fry resulting from unwanted spawning would be harmful in broodstock ponds.

very large ponds (0.5–1 ha) where there is less disturbance from the large constant throughflow of fresh water and a natural food supply in the early spring. In the spring the appetite of the selected broodstock increases as the water temperature rises. Although conditions in the small crowded wintering ponds are not ideal, many years of experience have shown that the carp can be well prepared for propagation in them. This is because the eggs were developed during the previous growing season. In spring they only need to reach suitable temperatures to spawn.

Fish farmers can easily make the mistake of feeding broodstock with large amounts of additional feed in the spring, as they do in the summer. However, this practice is more likely to be harmful than beneficial. This is because the ovary which is full of mature eggs does not have a need for additional nutrients. These nutrients can only therefore be used to develop fat, making the fish unsuitable for reproduction. This phenomenon is not accompanied by any external signs other than that the fish seem to be full of eggs but unable to reproduce. Consequently, in the spring, only protein-rich, energy-poor food should be fed to the broodstock.

During the period of rising water temperatures in the

spring, there is a danger of wild spawning in the broodstock when water temperatures reach 15–16°C. This can occur in the holding ponds releasing the eggs that were so painstakingly produced by the fish breeder. Such wild spawnings can be avoided if sexes are separated at the end of March while water temperatures are 10–12°C. When males and females are separated into wintering ponds it is possible that the sex determination of particular fish is uncertain (when based on external characteristics). In such cases, doubtful specimens should be transferred to the male pond, as any male accidentally introduced into the female pond could result in the wild spawning of the entire stock.

Sex determination is performed by observing the shape of the body and the relative position of the genital papillae (Fig. 3.9). Males will also release milt from the genital opening if carefully squeezed near the vent. Sex determination and selection is a good time for the breeder to have a final examination of the fishes' condition. If the stock quality looks poor then increased feeding can be applied. If they look fat, they can be starved for a period. This inspection also provides the opportunity to give the fish a quick bath to eliminate any possible parasites that might have developed during the wintering period. A breeding specialist is required to perform the autumn and spring selection as considerable experience is required to manage and select suitable broodstock.

When water temperatures exceed 18°C propagation can be

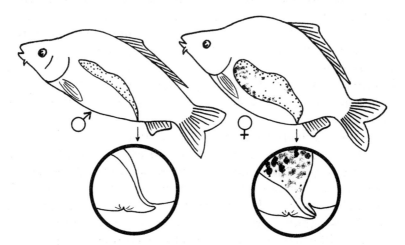

Fig. 3.9 Males and females can be distinguished by their shape and the appearance of the genital papillae.

planned. First, smaller females which appear most mature, i.e. with the softest bellies, are taken to the hatchery for the preliminary propagation. If these fish can be spawned successfully then the propagation programme can be scheduled.

3.4   Preparation of broodstock for artificial propagation

Broodstock prepared for propagation require particular care during harvesting selection and transportation.

The large ovary (15–20 per cent of the body weight) is contained within a thin layer of connective tissue rich with blood vessels. If broodstock are not handled carefully or the fish dropped, this membrane can rupture – a potentially serious injury that may result in the death of the fish a few days later. Even if serious injury is avoided, propagation could be prevented.

When harvesting, care should be taken to select the most mature or softest females first. Later the other fish can be used. Fish left in the pond should be kept in good condition and oxygen deficiencies should be avoided as should excessively muddy water which can coat the gills. Fish are best handled using a special net which is open at both ends. Fish are caught and released head first, which avoids damage or loss of scales (Fig. 3.10).

Fish should always be transported in water even over very short distances. This can be done using a double hammock of waterproof canvas attached to a solid frame. After filling the hammock with water, females are placed in pairs for direct transport to the hatchery (Fig. 3.11). Less physical work is required if fish selected for propagation are put into transport tanks on a lorry or trailer. Oxygen is bubbled into the water through a diffuser enabling 20–30 carp per cubic metre of water to be transported on a journey of several hours.

When bighead carp are transported, 10–12 ml of the anaesthetic quinaldine per cubic metre of water is used to keep the fish partially anaesthetised, i.e. in a 'calmer' condition.

Females are heavily anaesthetised inside the hatchery so that the weighing and initial hormone treatment can be performed without injury. It is advisable to commence this initial hormone treatment immediately to avoid additional handling and possible injury.

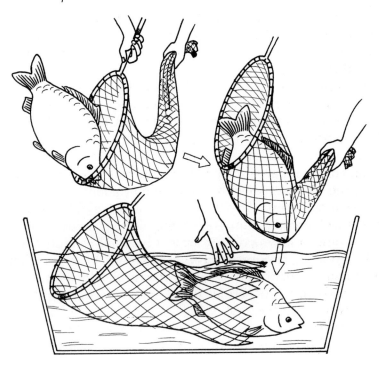

Fig. 3.10   Broodstock nets allow fish to be caught and released head first. (This avoids snagging the dorsal fin.)

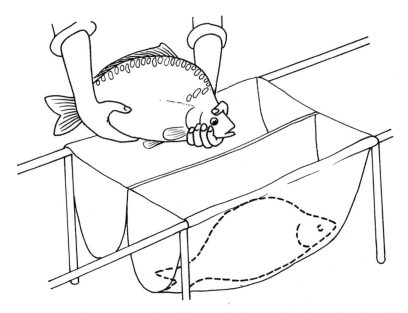

Fig. 3.11   The correct handling techniques for broodstock will avoid damage.

For anaesthetising (using MS 222 for example) a 1:10 000 dilution is used (10 g/100 l water). Anaesthetising should be performed in small tanks (100–150 l) to limit the amount of anaesthetic and hence the cost.

Six to eight fish can be anaesthetised together in 100 litres of water. This process is done on a continuous basis. Fish which have become narcotised are removed and replaced by fresh stock. Anaesthetising is complete when the fish are completely relaxed, i.e. can be turned over without resisting, and remain on their backs after they are released. Each fish should be in the solution for only a few minutes. If fish have been over-anaesthetised this is recognised by slow opercular movement, with little or no breathing (Fig. 3.12). If this occurs, fish should immediately be removed and placed into fresh flowing water, where normal breathing should return within a few minutes. During the injecting process, if there should be a delay involving a particular fish, then a worker should carefully watch to see that other fish do not become over-anaesthetised. Under these circumstances it could be safer to remove all fish back into fresh water.

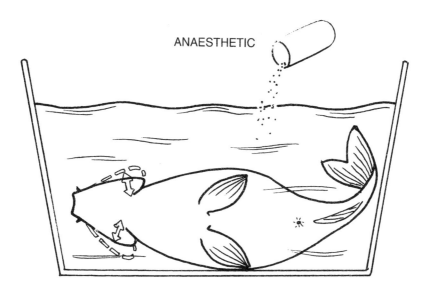

Fig. 3.12 If a fish has been over-anaesthetised the opercula will stop moving.

Using pituitary extract

At the start of the hypophysation technique, fish are weighed to enable the correct dose of pituitary to be calculated (according to body weight).

If the stock weight varies considerably, then fish should be marked by coloured threads tied in the dorsal fins. The 'key' to the colours should be recorded in a log book.

If it is possible to select a batch of fish of the same weight for treatment, this would be beneficial as far as calculations are concerned and can be easily performed with a little practice.

It is easier to work with batches of fish of the same weight. The weights of herbivorous fish are difficult to estimate and this can only be done with reasonable accuracy after considerable practice.

Once the weights of the females have been established the hormone solution is prepared. In the early days of using this technique fish were treated with one injection. However, it has since been established that the same dose given in two injections gives better results (Fig. 3.13). According to current knowledge 3.5–4.0 mg dry pituitary per kilogram of body weight is needed for egg release. Ten per cent of this should be given in a preliminary dose.

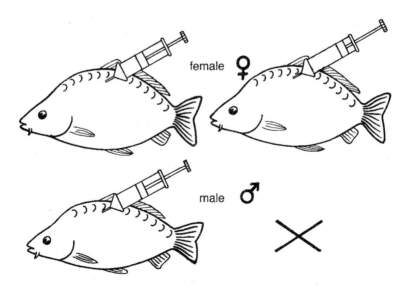

Fig. 3.13   The appropriate hormone dose is administered in *two* injections to the female, and *one* to the male.

As a preliminary dose, half a gland is given to a carp of 5–6 kg weight. Larger quantities than this are only given to very large fish, but less if during treatment of a batch fish start to release eggs after the initial injection.

> *Recipe for injecting carp:*
> (at a working temperature of 23°C)
> Time: 0 (preliminary injection): 0.3 mg/kg
> Time: 12 hours (decisive injection): 3.5 mg/kg
> Ovulation occurs 11–13 hours following the decisive injection.

To prepare the preliminary dose one dried gland is carefully ground into a fine powder in a mortar. The resultant powder is then well mixed with 1 ml of a 0.65 per cent saline solution. As some fluid is lost in the mortar during extraction 10 per cent more gland and solution should be calculated and made up for each batch (Fig. 3.14).

The grinding is more effective if a dry mortar is used and the glands ground into a fine powder. A few drops of saline are added and the mixture ground into a paste. The remaining

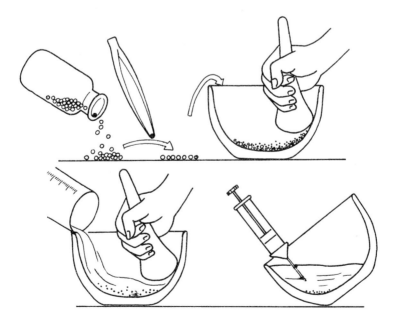

Fig. 3.14 Dried carp pituitaries are ground into a fine powder and dissolved in physiological saline.

saline is added, mixed and the solution then drawn up into a hypodermic needle and syringe. The solution is injected deeply into the muscle below the top of the dorsal fin using a medium size needle. The needle is kept in the muscle and a finger should be used to massage the area. This finger should then be placed over the entry point as the needle is withdrawn slowly to avoid losing any of the injected solution.

When 10–12 hours have elapsed (usually in the evening) the females are injected again. After anaesthetising, and before injecting, the genital opening should first be sutured to prevent eggs being lost from any wild spawning in the tanks.

The motionless fish is turned onto its back supported by a hand under its head. Strong cotton and a needle are used to produce an X-shaped stitch across the genital opening. The sutured females are then given the second (or decisive) dose of hormone whilst they are still narcotised (Fig. 3.15).

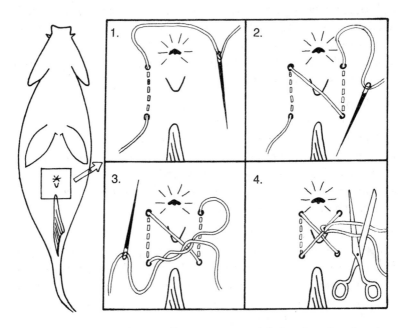

Fig. 3.15 The female papillae are sewn tightly closed using two simple stitches.

The glands are calculated to provide 3–3.5 mg/kg body weight. As before, they are ground in the mortar and diluted with physiological saline at a rate of 3 mg/0.2 ml solution, i.e. 1–2 ml of hormone solution (adding 10 per cent again for loss) for a 5–6 kg fish.

Males are treated once with 2 mg pituitary per kg body weight. The damaged (less valuable) glands are used for males.

After the hormone treatment, an undisturbed environment for the females must be provided (avoiding noise, lights, etc.) as stress at this stage will affect the time the eggs take to develop.

Using synthetic hormones

As an alternative to the hypophysation technique using pituitary extract, 'synthetic' hormones can now be successfully used. The basis of this method (known as the Lin-Pe method after its inventors) is that the synthetic hormone stimulates the fish to release its own sex hormones which control repro-duction. The artificial hormone is known as gonadotrophin releasing hormone (GnRH) and it is manufactured artificially in laboratories to work on a range of fish species. It is usually used in combination with neurotransmitters known as dopa-mine antagonists (such as metoclopramide, or domperidone). The use of GnRH is becoming very popular and is expected to replace pituitary extract because:

- the products are sterile (virus free);
- dosages can be calculated exactly;
- the injected material is not species specific;
- they should have less 'long term' effects on the sex glands of broodstock repeatedly spawned;
- they are cheaper and will ultimately be more freely avail-able than carp pituitaries.

The method of Gn RH treatment is provided in the instruc-tions accompanying the product. The products are sold in the form of a viscous liquid or in 'pellets', both of which contain all the necessary ingredients calculated for 1 kg of fish body weight. The liquid is injected 'neat' whereas the pellets are dissolved in water prior to use.

3.5   Stripping and fertilising the eggs

Egg maturation

As a result of the hormone treatment, physiological processes begin that stimulate the release of the matured eggs from the ovary wall and enable the fish to be stripped.

For this process to proceed it is essential to combine hormonal stimulation with specific environmental conditions. For the synchronised detachment of eggs it is necessary to provide water temperatures of 22–24°C in the tanks. If these temperatures cannot be maintained, maturation will be delayed or the eggs may be released spasmodically. The total amount of eggs may also be reduced. As well as a fluctuating temperature, low oxygen levels may also affect or even prevent maturation.

At 22–24°C the maturation process is complete 11–12 hours after the decisive hormone treatment. The exact determination of the stripping time is very important as early stripping prevents the process from completing, whilst late stripping results in over-ripe and badly fertilised eggs.

Ripening times

Temperature (°C)	Time (hours)
18	13–16
20	12–15
22	11–14
23	11–13
25	10–12

The exact time of stripping can be predicted by counting the hour grades, or by using indicator males. One or two males are put in with the treated females, and if left undisturbed they will start to spawn with the females (females cannot actually release their eggs because of the suture on the genital opening). 10–15 minutes after the first acts of spawning are observed, stripping can be commenced using the females which are splashing the most.

Stripping and fertilisation

For stripping, the fish are once again anaesthetised (the same solution can be used two or three times during a 24-hour period). Once anaesthetised, the fish are lifted out and placed onto the spawning table.

Here, fish have to be thoroughly dried with a clean towel, taking care that no water remains under the opercula or at the base of the fins. First, the suture is cut and the cotton carefully removed (Fig. 3.16). The right hand is used to massage the lower sides of the female from front to back, carefully releasing the ripe eggs from the ovaries into a plastic bowl (Figs 3.17, 3.18). The eggs should not be squirted forcibly into the bowl as they are sensitive to shocks. Stripping is continued until the ovary is empty or the eggs become bloody. In this case the female should immediately be placed into clean water. The stripping process should be completed as quickly as possible because, if the females remain in a deeply anaesthetised state for a long period of time, they may be severely harmed.

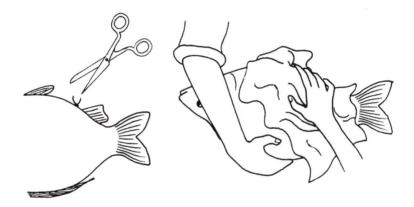

Fig. 3.16 Before stripping fish are dried using cloths, and the suture is cut.

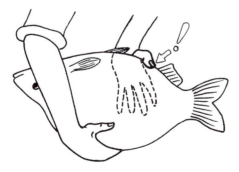

Fig. 3.17 The female vent is kept closed with the finger to prepare for stripping.

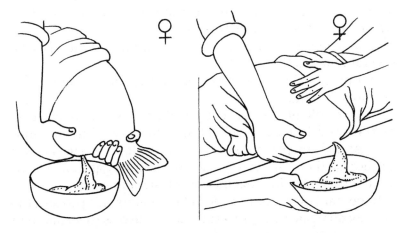

Fig. 3.18   Small fish can be held in the arms whereas larger females are laid on a table for stripping.

Stripped eggs are immediately weighed in their plastic bowl (which has previously been weighed empty). The egg weight is then written on the side of the bowl and in the spawning log book. Milt is then collected into small glass tubes from the anaesthetised males (Fig. 3.19).

The milt of two males should be used to mix with the eggs from one female. This should be done at a rate of one litre of eggs to $2 \times 0.5\,ml$ of milt. The milt and eggs are then thoroughly but gently mixed with a plastic spoon and then the fertilisation solution can be added. Fertilisation solution is

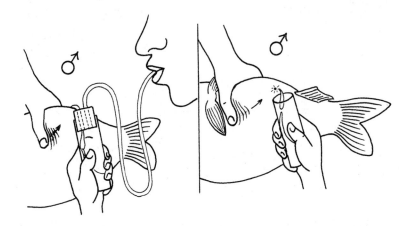

Fig. 3.19   Milt may be collected by sucking it into a pipette or squeezing straight into glass tubes.

made up using 40 g salt and 30 g urea (carbamide) in 10 l of water. After the fertilisation solution has been added to the eggs it is very important to stir the mixture continuously for several minutes (Fig. 3.20).

During this time the sperm penetrates the micropyle and fertilises the egg.

As a result of adding the fertilisation solution the eggs hydrate and begin to swell. During this period the swelling process and the presence of water activates molecules on the surface of the shell that cause stickiness. Eggs can be prevented from sticking together by constant mixing.

If too much solution is added at the start of this process, this also encourages the eggs to stick, so it is necessary to add only small quantities of the salt/urea solution. After, say, fifteen minutes the volume of eggs expands significantly and the first change of fluid should take place. The turbid supernatant liquid above the eggs is poured off and replaced with fresh solution. Also at this time, eggs are transferred from the 1–2 litre bowls into large 10–20 litre vessels. The swelling process lasts in general for $1-1\frac{1}{2}$ hours. After the first half hour the

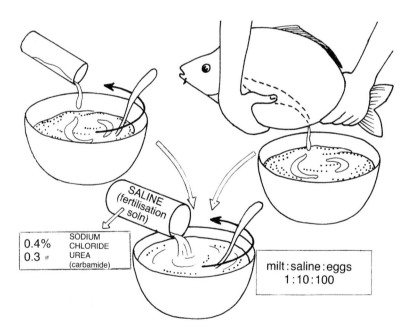

Fig. 3.20 Fertilisation can be performed using collected milt, or by stripping directly onto the eggs. After mixing, fertilisation solution is added.

stirring rate can be reduced but it is still necessary to replace the fluid (Fig. 3.21). Carp eggs will swell to six to nine times their original volume and the egg shells are quite resilient. When swelling is complete, eggs can be transferred into the Zuger jars. However, before this is done they should be treated with tannin solution to remove any final traces of stickiness (Fig. 3.22). Five grammes of tannin are added to 10 l

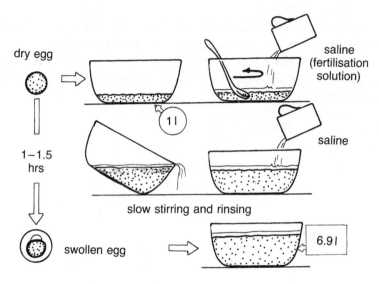

Fig. 3.21   The 'destickying' process using fertilisation solution takes up to $1\frac{1}{2}$ hours, until the egg stops swelling.

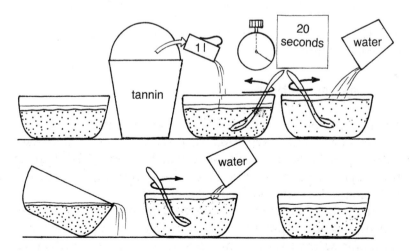

Fig. 3.22   The tannin treatment is performed rapidly.

of water and mixed. One litre of this solution is then added to 10–12 l of swollen eggs and mixed thoroughly by hand. The eggs are allowed to settle for a few seconds and then the tannin solution is poured off. The eggs are then washed several times with fresh water (which is the first time they have come into contact with water since their release from the female). The tannin solution is very toxic to the eggs. If they are kept in contact with this solution for more than 20 seconds they could be killed.

If eggs are still a little sticky after the first tannin treatment further treatments can be applied (usually two or three times at the most). After the final treatment the eggs are placed into the jars at a density of 1–1.5 litres of swollen eggs to each 7–9 litre jar (Fig. 3.23).

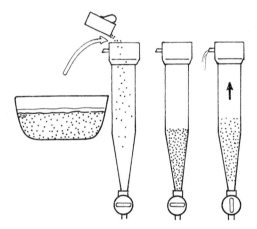

Fig. 3.23   Eggs are measured into the Zuger jars and the flow is immediately turned on.

3.6   Egg incubation

The development of the zygote commences immediately after fertilisation. With carp this process is quite rapid with embryogenesis lasting for $3–3\frac{1}{2}$ days at $23°C$ (Fig. 3.24). During this stage the eggs must be given the necessary care by providing suitable environmental conditions.

During incubation the oxygen demand of the eggs changes during the development stage. During the early stages when the proliferating cells are only loosely connected to each other, the egg is sensitive to mechanical damage (for 6–8 hours); thus

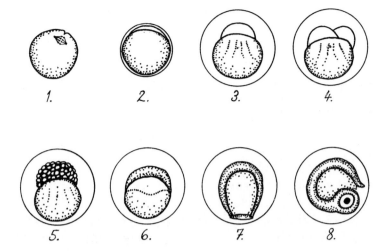

Fig. 3.24   Stages in the development of the embryo.
(1)   Fertilised egg.   (5)   Morula stage.
(2)   Cupola stage.   (6)   Blastula stage.
(3)   Swollen egg.   (7)   Gastrula stage.
(4)   Double cell stage.   (8)   Eyed egg.

the cell aggregate can easily be disrupted and destroyed. During this period oxygen demand is insignificant because the number of cells is small. At this time only a small volume of water is used to flow through the Zuger jars (0.5 l/min). As the embryo develops, the egg's metabolic rate increases and it requires more oxygen. The eggs start to release metabolites through the membrane and these need to be eliminated by the water flow. By the time the eggs reach the time of hatching the water volume flowing through the jars has increased considerably (2 l/min). During the incubation period, it is essential to monitor and control the flow of water through the jars (Fig. 3.25).

The development of the egg is signalled by a colour change from greenish-yellow, through brown to black, caused by the developing pigment cells (Fig. 3.26).

As well as controlling water flow, the eggs also require additional attention. Fertilisation rates may frequently be close to 100 per cent, which is far higher than could be achieved in the wild (say 40 per cent maximum). However, this still leaves a small number of eggs unfertilised. These eggs will die. Fungus, which will rapidly develop on any dead eggs, can then spread and kill healthy eggs if not

Fig. 3.25 Eggs should be monitored regularly throughout the 24-hour period.

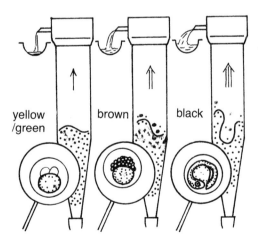

Fig. 3.26 The development of the embryo causes the eggs to change colour. At the same time their demand for oxygen increases, so the water flow is gradually turned up.

controlled. To prevent this, jars are treated with malachite green, which is the most effective and commonly used chemical fungicide. This chemical is effective at a dilution rate of 1:200 000 (5 ppm). If a stock solution is made up of 5 g/10 l, 10 mls of this stock is added to the Zuger jar. At the same time the water flow to the jar is switched off and the solution and eggs thoroughly mixed. After five minutes the original flow is returned and the chemical washed out of the jars (Fig. 3.27). Alternatively, a simpler method may be to add a more dilute treatment into the header tank which feeds all of the jars. This is a longer duration treatment (over, say, 30–60 minutes) and should be at a concentration of 1:500 000 (2 ppm). Other methods can also be used to prevent dead eggs from causing problems. One and a half to two days after fertilisation the inner part of the unfertilised egg disintegrates and the specific gravity of the egg changes. As these dead eggs are lighter than the live eggs, the water flow causes them to congregate on the top of the egg mass. If the fertilisation process has been efficient then only a relatively small number of dead eggs will be present, but if the

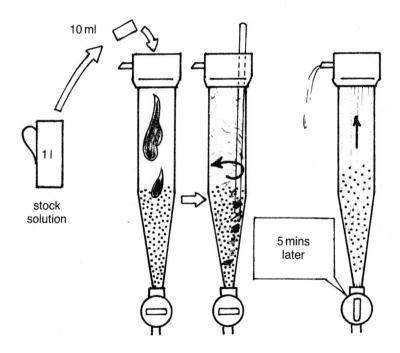

10 ml

1 l

stock
solution

5 mins
later

Fig. 3.27   Malachite green treatments will prevent the development of the fungus *Saprolegnia*.

number is large these dead eggs can form a layer several centimetres thick, posing a constant threat of fungal infection. The hatchery manager must therefore eliminate this layer periodically (Fig. 3.28). The dead eggs can be siphoned off using a thin tube of 10–15 mm diameter, taking care not to siphon any live eggs.

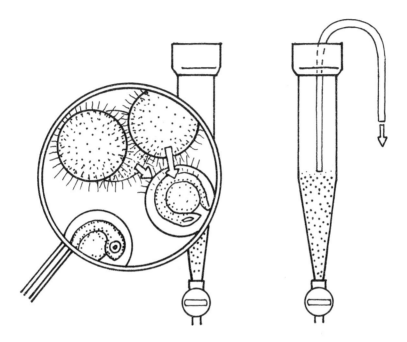

Fig. 3.28 The layer of dead eggs should be removed to prevent fungal contamination.

Near hatching time the developing embryos, now with a very noticeable black eye pigment, can be seen moving more and more within the egg. The actual hatching process is stimulated by this mechanical friction as well as by the production of a 'hatching enzyme'.

3.7   Control of the hatching process

If the oxygen content of the water falls, the activity of the embryo increases as it attempts to escape from the unfavourable environment. Therefore, by decreasing water flow (thus causing oxygen levels to fall) the hatching process can be stimulated. This technique should only be used if the

natural hatching process has started, otherwise larvae will
hatch out prematurely.

If the water flow to the jar is reduced after the hatching
process has commenced, hatching can be synchronised and
speeded up. This may be important if jars are required for
the next batch of eggs. The technique involves reducing the
flow for five or ten minutes until the surface of the egg mass
is hardly moving. After this period the flow is increased
slightly and a few minutes later the rest of the eggs will
hatch simultaneously and the mass can be siphoned out of
the jars.

The hatching eggs are siphoned through a tube into a flat-
bottomed vessel with a large surface area (Fig. 3.29). This
process should be performed with considerable care, mini-
mising the velocity at which the eggs are siphoned in.

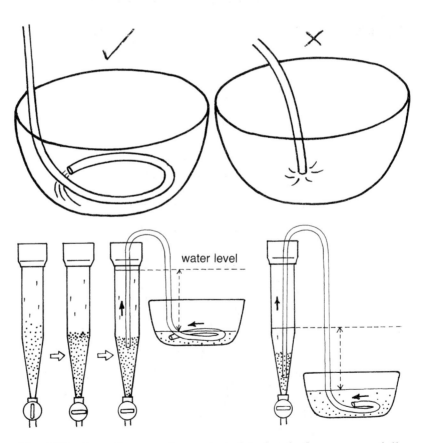

Fig. 3.29 Once the eggs have started to hatch they are carefully
syphoned into a hatching bowl.

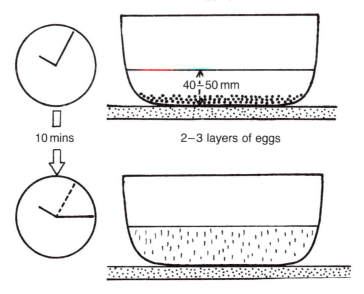

40±50 mm

10 mins   2−3 layers of eggs

Fig. 3.30 Hatched larvae attach to the wall of the bowl.

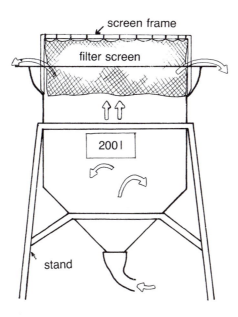

screen frame

filter screen

200 l

stand

Fig. 3.31 The 'larvae container' used for larvae rearing.

Larvae and eggs are mixed in the bowls. After a few minutes the increased movement of the embryos, plus the naturally-produced hatching enzyme (which is acting on the egg shell) physically disrupt the egg shells and the larvae hatch (Fig. 3.30).

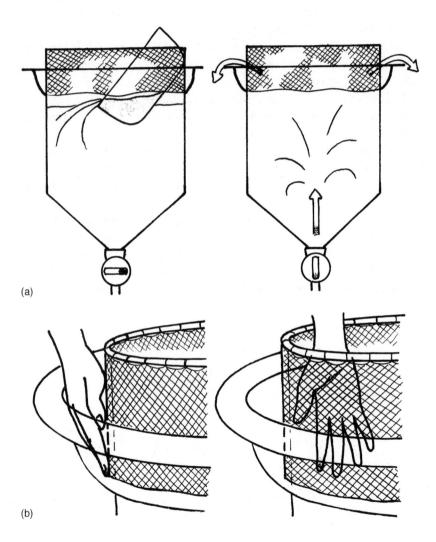

(a)

(b)

Fig. 3.32 (a) The flow should be started immediately larvae are introduced into the incubator. (b) The screen should be cleaned frequently (outside surface first).

3.8   Larval rearing in the hatchery

The hatched larvae, which as yet cannot swim, are transferred from the bowls into the large 'larvae containers' used for larvae rearing. Half a million larvae can be transferred into a 200 l vessel (Fig. 3.31). In this container most of the larvae will gradually work their way up to the surface, where there is a mesh ring, whilst the remainder stay turning over in the throughflow of water entering from the bottom. During this period, husbandry involves the regulation of water flow and the continuous cleaning of the mesh (Fig. 3.32).

Hatched larvae have developed enough to be able to swim in water temperatures of 20–24°C after three or four days. Swimming is only possible once they have filled their swim bladders with air. At the same time, their digestive system is now able to receive and digest external food. This period marks the end of the hatchery phase. Further development and rearing of the feeding larvae is more secure in a pond environment. Simultaneously the fry ponds have been prepared to receive the feeding larvae whilst they receive their initial feeding in the hatchery.

Once they are physiologically ready to feed, the larvae immediately seek external foodstuffs. The simplest feed to provide is finely ground hard-boiled egg yolk, mixed in 0.5 l water (Fig. 3.33). This is fed at, say, one egg per 100 000 fry per day for one or two days at the most. The egg breaks up into

Fig. 3.33 Hard-boiled eggs are finely ground in a blender and mixed with water.

Fig. 3.34 The egg particles are fed to the larvae regularly throughout the first day or two of feeding.

particles of a few hundred microns in diameter, which can be easily consumed by the larvae (Fig. 3.34). Unfortunately this is not a complete diet and can only be used as a feed for a couple of days prior to transferring them into the pond, which is a much better environment for providing nourishment.

Chapter 4
The Technology of Fry Rearing

4.1 Selection and preparation of ponds

The requirements for fry rearing ponds can be satisfied only by those ponds which are built specially for this function (Fig. 4.1). The most important requirements of the ponds are:

- they should not be built too large (a maximum area of 5000 m^2);
- excellent quality water supply;
- perfect drainage;
- good vehicle access;
- suitable harvesting devices (e.g. designed for small fish).

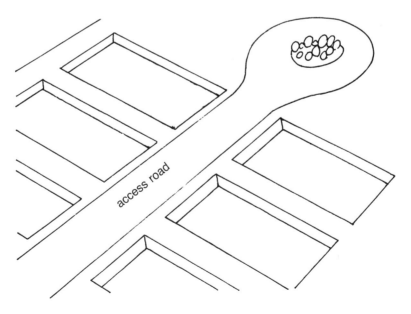

Fig. 4.1   A bank of nursery ponds with common road access.

Ponds not specially designed for fry rearing may also be used for the purpose but results will not be as good.

The techniques of pond preparation

The reason for preparing advanced fry rearing ponds is to enhance fry survival. Final preparation is started two weeks before stocking is planned (which is in addition to the autumn to spring maintenance work and includes the drying process which influences pond production).

Preparation starts with the cleaning of the pond bottom and the spreading of quicklime (Fig. 4.2). This is to eliminate conditions that may encourage the survival of disease-causing agents. After cleaning, the ponds are flooded. If there is a possibility that wild fish may be introduced from the inlet, water should be passed through a filter box containing a mosquito net (Fig. 4.3). As such a small mesh may be quickly blocked, it should be frequently cleaned.

Ponds are filled up to the half or two-thirds working level. The water is then fertilised with both organic and inorganic fertilisers. Superphosphate is added at 100–150 kg/ha. In

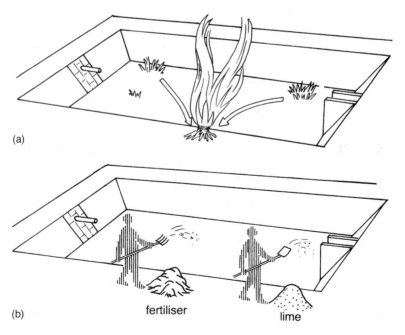

(a)

(b)   fertiliser   lime

Fig. 4.2   (a) Weeds must be removed annually from the ponds. (b) The pond bottom is then lined and fertilised.

Fig. 4.3 A basket (box) with a fine mesh screen prevents the entry of predatory fish and insect larvae.

ponds without fish, organic fertiliser can be used at a high concentration as fish stocks will not be immediately endangered. Ripe manures should be added at 3–7 tonnes/ha, which is higher than for ongrowing ponds as fry ponds do not receive maintenance fertilisers later in the season. Nitrogen fertiliser is added at the inlet, where it will immediately dissolve, whereas superphosphate and organic manures are spread from a boat.

Fertilising creates favourable conditions for algae and lower crustaceans, and hence these planktonic organisms will multiply very quickly. Since rotifers are the most favourable organisms for early fry to feed upon, the pond is additionally managed to select for these organisms. This is done by adding particular chemical treatments that selectively kill other groups of zooplankton which themselves feed on or compete with the rotifers and prevent them from developing quickly (Fig. 4.4). Certain organophosphorus

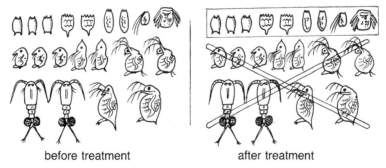

before treatment   after treatment

Fig. 4.4 As a result of the insecticide treatment the plankton population within the pond changes composition. The copepods and cladocerans are removed leaving the rotifer population to develop.

insecticides (Dichlorvos or Dipterex etc.) are suitable for destroying these harmful invertebrates. Dichlorvos should be used at 1 mg/l (active ingredient of 'Trichlorfon' = 40% w/v). Care should be taken as the strength of the insecticide seems to vary between products, and thus an experimental treatment should be first tried. If water is treated at 1 mg/l (i.e. 1 g/m^3), these harmful invertebrates are killed within 24 hours. The amount of chemical required is calculated and dissolved in a bucket with pond water. If the pond is small, the solution is evenly applied around the banks of the pond (taking great care to avoid skin contact as the concentrated chemical is poisonous). CHEMICAL TREATMENT SHOULD ONLY BE DONE BY SPECIALLY TRAINED PERSONNEL. In the case of larger ponds, a boat should be used with one person steering and the other sprinkling the chemical into the water. (Fig. 4.5).

Fig. 4.5   Chemical treatments should only be performed by trained personnel wearing protective clothing.

Favourable results can be seen within a few days, with first the algae and then the rotifers multiplying vigorously. This can be quantified by taking plankton samples.

During a plankton check, 100 l of water is filtered through a plankton net and this is poured into a graduated glass tube. The filtered organisms are killed using a solution of formalin and allowed to settle for a few hours prior to reading (Fig. 4.6). This information is a useful parameter for use in fish husbandry. If there are 0.5–2 ml of plankton/100 ml water the rotifers in the pond are at an acceptable density.

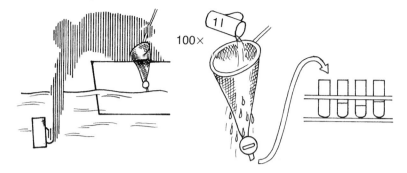

Fig. 4.6 To measure the density of the plankton population, 100 litres of pond water is filtered and examined.

4.2 Transport and stocking of feeding fry (larvae)

The timing of pond preparation

The preparation of a pond commences at the time when broodfish are brought into the hatchery. Large ponds may be started a few days earlier to ensure that the rotifer population reaches the required density.

In the hatchery, whilst larvae are hatching and starting to feed, the rotifer population is developing in the pond. This population should be at optimum density (hence optimum feeding conditions for the fry) to coincide with the time when the fry are ready to stock out.

Husbandry prior to stocking

Before stocking, fish are fed several times in the hatchery with a diet of hard-boiled egg. This feed does not totally satisfy the nutritional requirements of the fry but it is excellent for teaching the fry how to feed, and provides a certain amount of energy which may quickly become limiting. Energy reserves in the fry are small, and the fish farmer must therefore stock out the feeding fry at the correct time.

During stocking, care should be taken to ensure that fish are not subjected either to rapid changes in the environment or to physical injury. Fry are usually stocked early in the morning but if ponds are large and subjected to strong winds or rapidly decreasing temperatures then the stocking should be postponed.

The postponement should last only one or two days at the most as there is no available feed for the fry. The hard-boiled egg is nutritionally insufficient and it would not be practicable to try and collect sufficient rotifers from the prepared ponds. In such cases it is better to sacrifice the stock of fry and start again with a new propagation of broodstock. This option may seem drastic but it is preferable because if only a small percentage of fry survive the bad weather they will consume any newly stocked fry added to the pond to replace the losses.

Transport methods

Fry transport must be carefully planned. If ponds are close to the hatchery fry can be transferred in buckets or bins (e.g. 100 000 per 10 litres for, say, five to ten minutes) (Figs 4.7, 4.8). Many million fry can therefore be stocked on the farm by this

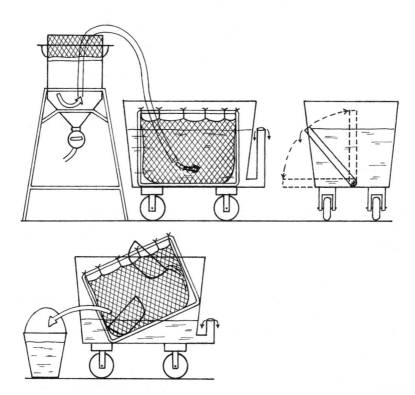

Fig. 4.7   Larvae ready to be transferred into the pond are syphoned into a net cage. Excess water can be poured away and the larvae can then be removed with a bowl.

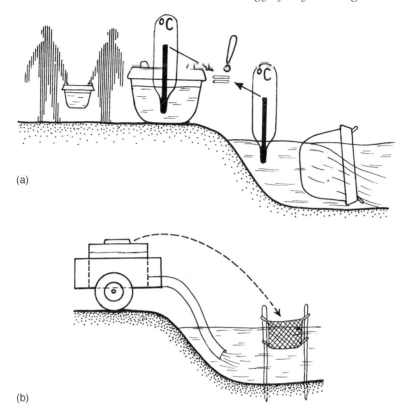

(a)

(b)

Fig. 4.8   (a) Small numbers of fry can be restocked using bins. Care should be taken to balance water temperatures. (b) Large numbers of fry are transported in tanks. A trial cage should be stocked with a few hundred fry so that their progress can be easily monitored.

method. For longer journeys specially fabricated glass fibre transport tanks, 1–1.5 m^3 in volume, may be used supplied with a finely diffused oxygen supply. As many as 1–1.5 million fry can be transported in such tanks for journeys lasting several hours.

The fry are released into the ponds either by syphoning through a flexible hose or by using a purpose-built plastic chute, which attaches to the tank. The tanks themselves are transported on trailers or special trucks.

Fry are the easiest of all age groups to transport. Their oxygen demand is very small, as is the volume of water required to carry them. The oxygen input only needs to be slight as vigorous bubbling would merely exhaust the fry. During transportation periods exceeding 2–3 hours the fry will empty its gut. Therefore it is worth taking a supply of prepared egg mix to feed to the fry when the fish are checked every hour.

For extremely long journeys, particularly in the case of smaller numbers of fish, plastic bags may also be used. 100–150 000 fry can be transported for several days in a 20-litre bag pressurised with oxygen (one-third water to two-thirds oxygen) (Fig. 4.9). Whatever method is used to transport the fry, it is essential to equalise the temperature of transport water gradually with the receiving water. This may be done by floating the plastic bag in the pond for, say, 30 minutes, or when transport tanks are used, a quantity of pond water is

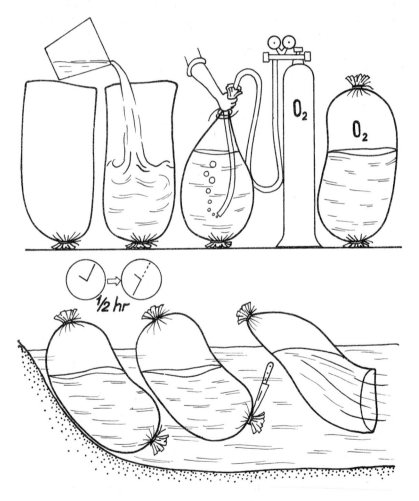

Fig. 4.9 Fry that require transporting over very long distances should be stocked from polythene bags. Allow time for temperatures to equilibrate.

gradually poured into the transport tank until temperatures are equal. This procedure is essential for the receiving farmer to ensure successful stocking.

There should be no more than 1–2 °C difference between the two waters.

At the time of release a sample of the fry should be held in an 'observation tank'. This 'observation tank' is a fine mesh cage which is held in place near the edge of the pond. A few hundred fry are placed in this cage which can then be observed throughout the following weeks. If all is well these fry should thrive, but if there are problems with the prepared pond or with the fry themselves, then these fish may start to die. This could be due to poor viability of the stock as well as the presence of cyclops in the pond or high pH due to over-fertilisation. If this fry sample is seen to perish then the fry in the pond may die also.

If problems are recognised early then the production loss can be rectified by introducing new stock, improving the water quality or by leaving the particular pond out of the production cycle (using a fresh pond instead). However, the reason for fry losses should always be identified in each case.

During pond preparation, a good fish farmer will check the plankton population several times to ensure it is developing well. This checking should continue during and immediately after stocking. If the rotifer population on the day of stocking is too low, it unfortunately cannot be increased. If, however, there is evidence of cyclops appearing it would be beneficial to perform a second chemical treatment to kill them off and thus prevent higher fry losses.

4.3   The development of the plankton population

If stocking is performed in good weather, the fry quickly become used to the new environment. After a few hours they may be found in the corners and around the edges of the pond. A few days after stocking the survival rate of the fry may be determined. Fry are usually stocked at 200–$400/m^2$ or even as high as $600/m^2$ in extreme cases (depending on how early they are to be harvested). Survival will be determined by the amount of nourishment that the fish receives as well as the chemical and physical characteristics of the pond environment.

By preparing the pond for the development of the rotifer population the most dangerous period for the fry will be overcome, but a total food supply is not achieved by this. Rotifers can only satisfy the food demands of quickly growing fry for a few days. After this, larger food organisms should be available to satisfy the nutritional requirements of the fish. To promote the development of these larger planktonic organisms, the fish farmer has to manipulate the biological processes at an early stage.

The control of the plankton population

The medium-sized zooplankton will perish as a result of the chemical treatment. The natural restoration of their population would normally take several weeks, so the fish farmer has to accelerate this process by inoculating the ponds with the correct organisms five days after the decomposition of the chemical.

In warm summer periods the most rapidly multiplying organisms are *Moina*. The pond should be inoculated with this medium-sized crustacean at the same time as the fry are stocked. One or two buckets of *Moina* (i.e. a few hundred thousand) are enough for a pond. Following this inoculation the fish fry are far too small to catch and feed on the *Moina* so they can multiply freely, becoming an ideal food for the fish 10–12 days after stocking (Fig. 4.10).

If it is not possible to inoculate plankton, a 'natural' restocking can be stimulated by further flooding of the pond after stocking the fish. Unfortunately, not even these medium-sized plankton can satisfy the increasing nutrient demand of the fish until the end of the fry rearing stage. A third plankton stage can therefore be included to provide a population of very large food organisms. This is done by placing a few buckets of *Daphnia* into the rearing pond 10–12 days after stocking the fish. They will multiply within 8–10 days and provide abundant natural food for the last, third stage of rearing.

Fry nutrition

According to present knowledge, the young fry possesses an underdeveloped digestion system which is best served by obtaining nutrients from living organisms. Thus it is essential

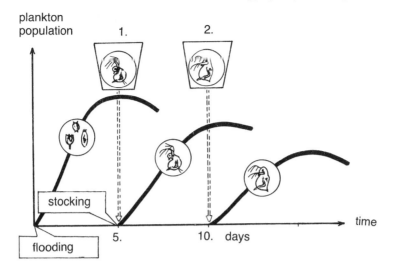

Fig. 4.10 Insecticide treatment and plankton inoculation ensures the development of the most favourable types of natural food: (1) *Moina* spp; (2) *Daphnia* spp.

to ensure the presence of very large numbers of food organisms of the appropriate size to provide adequate nutrition (Fig. 4.11). Later, as the digestive system develops, the fry can consume mixed feed, including appropriate sizes of fine artificial feeds known as 'flours'. This feed should be supplementary to the live feed and should not form a complete diet. Feeding should be commenced on the day of stocking. Although the fry will not feed on the flour immediately, it will not be wasted as the planktonic organisms will consume the finest grains. Hence, initially, only the plankton is being fed. Simultaneously, however, the fish are becoming accustomed to the taste of the feed and become used to a mixed diet.

As the fish grow, the artificial feed becomes increasingly important until the stage is reached when it is virtually the sole diet (Fig. 4.12). Because of this the quality and content of the feed is very important. It is particularly important at the early stages to provide proteins originating from animal sources and soya.

The following mixture is well tried and tested:

25 per cent barley or wheat flour
25 per cent soya flour
25 per cent fish meal flour
25 per cent meat or blood flour

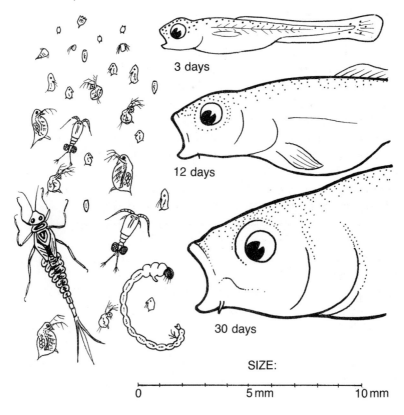

Fig. 4.11   A complete development of the natural pond fauna ensures a continuous supply of the correct size food for the growing fish.

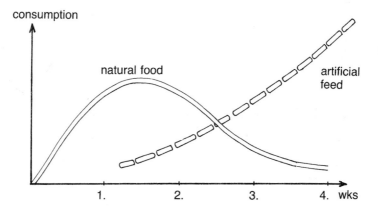

Fig. 4.12   The food consumed by the fry changes during the first weeks of nursing.

These 'home made' diets are significantly less expensive than commercially produced products. The feed rate for these flours should be 1–1.5 litres/day per 100 000 stocked fish Later, the quantity will be modified according to survival of fish and plankton concentration and will ultimately be increased to four or five times initial feed rates.

The feeding can be performed in several ways – for example, cast onto the surface of the pond (Fig. 4.13) or wetted and poured in. In the early stages the latter method is more effective but as the fish grow they can graze the feed off the surface. During the rearing of the fry, the growth and survival of the stock should be regularly checked (Fig. 4.14). In the early stages, survival of the advanced fry can be appraised by observing the grassy pond margins. A white plate can be sunk under the surface and the dark fry can be seen against this background. Later, hand nets should be used to sweep through the margins to catch a few fry. By the end of the fry rearing stage, fish may only be sampled using a throwing net. Fry should be placed into a glass beaker so that they can easily be observed for abnormalities in size or behaviour. If there is any evidence of abnormality a sample of the stock should be sent for microscopic examination. Occasionally, parasites may develop on the stock and these can be identified by the

Fig. 4.13 In the days following stocking, powdered food is fed to the fry.

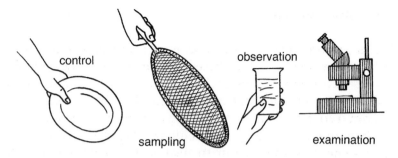

Fig. 4.14   The development and health of the stock are continuously monitored during the fry rearing phase.

biologist or veterinary surgeon who can advise on the correct treatment.

At the end of the fry rearing stage the environment conditions deteriorate as all of the natural food has been consumed as the fry population has grown. The fry rearing stage is thus complete and they are now ready for harvest.

4.4   Two-step rearing

As stated earlier, this involves harvesting the 'advanced fry' after 4–5 weeks to restock into another pond. The best time to harvest fry depends on the working practice of the farm. If temperatures remain between 20–25°C and there is sufficient food, the fish can be harvested three or four weeks after stocking. Fry survival will be high compared with stocks harvested at the six to eight week period (the difference observed may be as high as 20–40 per cent). The longer the fry rearing period lasts, the larger the fish will grow. Three-week-old fry will weigh 0.1–2 g whereas six- to eight-week-old fish may reach 1–2 g. Whether large numbers of small fry or smaller numbers of large fry are beneficial depends on the working practice of the farm.

Recent practices have led to the development of large fry rearing ponds (bigger than 2 hectares). It has been shown that fry growth rates can be faster in these larger ponds and harvesting generally commences after three to four weeks. At this time the fish weight will be approximately 0.5 g. As large ponds take longer to harvest fry continue to grow as the harvesting and draining is started. By the time the pond is finally emptied the fry have grown to a considerable size.

Harvesting methods

For harvesting advanced fry, seine nets are used made of a very fine mesh (2–3 mm). These nets are at least 15–25 m long, and to fish pond depths of 1.3–1.5 m, material of 2–2.5 m is used. Seine nets are floated on the top line with a weighted bottom. Wooden poles on either edge of the net are hauled by the fishermen at the appropriate depth (Fig. 4.15). To harvest with seine nets the fisherman nearest the bank stands still while the other sweeps around towards the bank. Once the bank is reached the lower rope is tightened and pulled out-wards. Fry are shaken down into the middle of the net from where they are hand-netted out (Fig. 4.16).

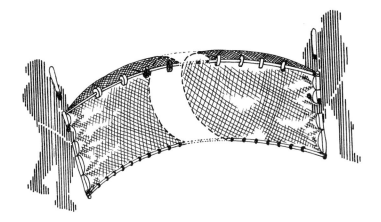

Fig. 4.15   A fine seine net is used to harvest 'advanced fry'.

Seine netting is an indispensable method for harvesting large ponds and is also useful for small ponds, particularly if the water level has been dropped by half.

By removing a high proportion of fish by this method fry are damaged less than if the whole population were harvested at drainage. When large numbers of fish are crammed into a harvesting device at the back of the pond losses may quickly accumulate.

For catching fish at drainage there are several kinds of harvesting device. The most popular is the catching box fixed on the outlet pipe, the bottom of which is solid (e.g. made of wood) whilst the sides are covered by a strong mesh netting. Another type is composed of a net 'box' containing a V- or

Fig. 4.16   The net is operated by two competent workers.

funnel-shaped entrance. This prevents fish from swimming back against the current. Several of these can be placed side by side on large outlets (Fig. 4.17).

The harvested fry are taken to fattening ponds and stocked at a low density. If these ponds are close to the fry ponds the fry are taken directly on a transporter vehicle and transferred to the new pond.

If the fish need to be taken to ponds more distant from the

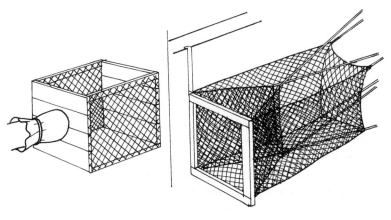

Fig. 4.17   Equipment used in draining the pond includes fry trap boxes and trap nets.

fry rearing ponds, the fry should be temporarily stored. This is best done in large concrete or fibreglass tanks with a constant water supply. These temporary storage ponds can contain large cage nets for ease of handling (Fig. 4.18). Storage of the fry is also necessary to allow the fish to empty their gut, an essential requirement prior to transportation. During this holding period damaged fish will die. The behaviour, condition and health of the stock can also be assessed. When enough fish have been collected they can be moved to the new pond in the coolest time of the day (i.e. at dawn).

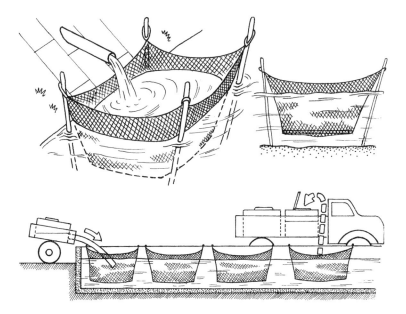

Fig. 4.18   Fish are released into holding nets from where they are easily removed for transportation.

Prior to transporting, the harvested fish can be counted. Counting fish can be done according to weight or volume. Bigger fish are best counted by sample weight whereas smaller, more sensitive fish are best counted by volume. Both methods involve counting a number of individuals in a known volume or weight, which is then compared to the total quantity or weight of the total stock (Fig. 4.19). A simple calculation is used to determine the total numbers of fish present. Both methods are satisfactory and should have an error less than 5–10 per cent. To reduce the margin of error still further, larger sample volumes or weights should be counted.

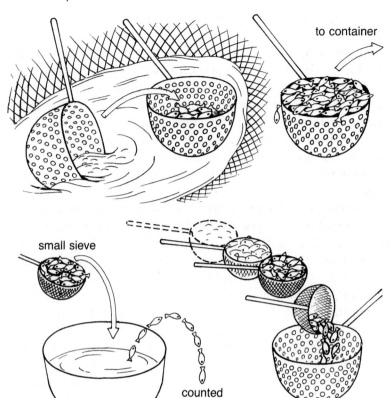

to container

small sieve

counted
sample

large sieve

Fig. 4.19   The simplest and most commonly used method for estimating numbers of advanced fry is to count a known volume.

4.5   Single-step rearing

One-summer fish can be produced in two ways:

- by harvesting the advanced fry after 4 weeks to restock and
  ongrow for the rest of the summer in another pond: the
  two-step method (described earlier in this chapter, section
  4.4)
- or in one single step by stocking larvae and leaving them to
  grow in the pond for a whole season.

The process of single-step rearing involves higher risk and
requires careful and constant monitoring. For this rearing
programme, pond preparation is carried out following the
same principles that were described earlier. However, preventing wild fish entering the pond is essential or these

unwanted species will be growing and competing with newly introduced fry for a whole season.

The starting point of single-step rearing involves fry production in the hatchery just as for two-step technology. The stocking of fry into the pond is carried out according to normal practices. In addition, a sample of several hundred fry can be introduced into a small fine-meshed cage anchored in the pond, where the behaviour as well as the survival of the sample can be monitored after a couple of days. This will demonstrate the vigour of the fish and the quality of the pond water.

Traditionally, large ponds (2–3 hectares) are used for this type of culture. The ponds should *not* be overstocked because the survival of fry can only be high when the favourable starter food (rotifers) is available in large densities. In two-step fingerling rearing the number of fry stocked in 1 hectare can be as high as 1–5 million, but in the single-step there should be only a fraction of that number, say 1–200 000 fish/ha. Since on larger lakes the wind can generate waves that are harmful for the fry, it is necessary to create hiding places for the fish on the shallow margins of the pond by cutting grass and marsh vegetation on to the water surface. This floating material also functions later, as it attracts mosquitoes and midges to lay their eggs, and the resultant larvae of these insects provide a valuable natural food for the fish over the summer. Healthy and active fry can be observed in a couple of hours after stocking, even in the remote parts of the pond as they quickly spread and find optimal habitats for themselves.

The fry should be fed with 'flours' (fine mixed feeds) introduced around the margins of the pond. Later the feeding can be concentrated to certain areas in the pond, which will subsequently become the feeding points.

In single-step fry rearing technology, grass carp, silver carp, bighead carp and catfish can also be reared together with the carp fry.

The most important problem of single-step fry rearing is the control and management of aquatic vegetation. As fry are so small, they are unable to 'turn up' the bottom silt layer, which would cloud the water and promote algal growth. The pond therefore remains clear, sunlight penetrates to the bottom, and aquatic vegetation starts to grow vigorously. Plant growth can be very prolific even if older age groups of fish were reared in the pond in the previous year, as seeds, rhizomes, overwintering shoots, etc. remain from earlier seasons that quickly develop into new vegetation.

The harmful effect of water plants is many-fold:

- Plants efficiently absorb the nutrients dissolved in the water, thus preventing the development of algae and bacteria which serve as food for zooplankton. The crystal-clear water of ponds rich in aquatic vegetation is a well-known problem and a sure sign of the absence of planktonic life.
- Aquatic plants harbour the larvae of many predator insect species, which hide in the vegetation. These insects can cause incredibly high losses in the fry stock by their predatory behaviour.
- A large biomass of plants disrupts the biological stability of the pond, particularly the oxygen balance. The volume of dissolved oxygen is very high and carbon dioxide absorption is very efficient during the daytime, but at night plants absorb oxygen, thus having the same effect as a dense bloom of algae.

Cutting tools such as a scythe should therefore be used in these cases to cut out the vegetation. The harvested material should be allowed to blow to one end of the pond where it can be removed. The desired muddiness of water can be achieved by dragging heavy metallic objects (harrows) along the pond bottom.

As with all gradually overpopulating production systems, a lack of protein can occur towards the end of summer if the survival rate of the fry has been high. In such cases pelleted food can be fed to supplement the natural food, following calculations based on plankton abundance and food demands of the predicted fish stock.

Further technological details of single-step fish rearing are identical to the ones described under two-step rearing (section 4.4).

Chapter 5
The Annual Cycle of Carp Husbandry: Ongrowing Fingerling Fish to Market Size

5.1 Pond preparation

To enhance the ability of ponds to generate natural food it is beneficial to keep the pond dry for a period of time throughout the year. During this time, organic material can break down and pathogenic bacteria and parasites will die off. Escapee fish will also be eliminated, and the soil in the pond bottom regenerates ('mineralises').

It is very important to dry out the pond in order to ensure the success of the next year's production. If the pond remains under water for several years, the biological production decreases significantly. Fish diseases become more frequent and the biological condition of the pond becomes unstable. This temporary drying out of the pond is therefore considered to be essential in enhancing overall farm production.

After draining, ponds should be kept dry to allow time for pond repair as well as to increase the production of the pond bottom. The repair of dams, inlets and outlets (monks) should be treated as a priority.

Water leaking through even small holes in dams can rapidly enlarge them, leading to the possibility of further damage and even bursting dams. Holes created by voles and other pests should therefore be identified and repaird (Fig. 5.1). Wave action can undermine banks and dams. It is common to repair large areas of damage with woven branches to act as a defence against further damage (Fig. 5.2).

If reed or other bankside vegetation has encroached into the pond, it is usually easier to cut and remove when the pond is dry. The vegetation can be burnt *in situ* along the bank side.

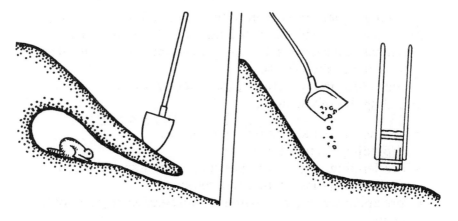

Fig. 5.1 Holes made by bank voles have to be repaired to avoid damage to the dam.

Fig. 5.2   After draining stakes can be used to prevent wave damage.

Sterilisation and improvement of the pond bottom

The main aim of pond preparation in the spring is to sterilise and improve the quality of the pond soil using lime. This can be performed with conventional equipment or, for large areas, liming may be done from aircraft. If the soil is hard enough heavy machinery plant can be used to harrow or rotavate the upper layers. Nets and other muddy surfaces should also be sterilised.

Liming is performed before fertilisation to ensure that the pH of the water will be sufficiently high to allow mineralisation to take place efficiently (i.e. pH 7–8) enabling the fertiliser to have the desired effect.

Lime has many effects upon the pond environment, including the following:

- raising the pH and buffering capacity;
- providing calcium for invertebrates;
- accelerating the decomposition and mineralisation of organic material, particularly cellulose;
- precipitating organic matter in suspension (e.g. colloidal clay);
- acting as a disinfectant (when used in certain forms).

Lime is available in three forms: quicklime, slaked lime and limestone.

(1) *Quicklime (burnt lime) (CaO)*
This form is the most reactive and hence is dangerous to use. Because of this it is not freely available in some countries. It reacts vigorously with water to form calcium hydroxide, generating much heat in the process.
It is used at 500–750 kg/ha in thick mud or 200–300 kg/ha in thin mud.
It also has strong disinfectant properties (at 1000–1500 kg/ha).
(2) *Hydrated lime (slaked lime) ($Ca(OH)_2$)*
This form is sold in bags as builders' lime. It is quite reactive and thus should be handled with care, but is more freely available than quicklime.
It is used at 1.5 times the rate for quicklime and is used for the same functions.
(3) *Limestone or chalk ($CaCO_3$)*
This is an inexpensive form of calcium if the appropriate rock types are found in the locality. It is quarried and usually sold 'crushed'. This form of lime does not have disinfectant properties and should not be used in muddy ponds. It is used at 750–1000 kg/ha.

High doses of nitrogen (in the form of ammonium nitrate) can also be spread over the pond bottom in high concentrations to kill wild fish. Later, during the flooding of the pond, it is diluted and becomes a good nutrient source.

Flooding the pond

After preparing the pond bottom it can be flooded. This usually happens in early spring and should be carried out with considerable care.

Ponds should be filled by passing the water through a mesh or grating small enough to prevent wild fish from entering. Ponds for marketable size fish are also flooded at this time. Filter screens should be cleaned regularly to avoid a reduction in the rate of flow otherwise flooding may take much longer.

5.2   Fertilising and natural food development

The aim of additional fertilising is to provide an environment where plant and animal organisms can grow and reproduce quickly. This ensures that sufficient natural organisms are available as food for the young fry. The cheapest form of producing fish 'meat' is to provide optimal levels of nutrients in the pond water. The most effective and most common way of providing these nutrients is by fertilising with organic and inorganic fertilisers. Pig, poultry and cow manures are the most commonly used forms of organic fertiliser, due to their low cost and ready availability. (Sheep, rabbit and horse manure are less commonly used.) Details of application rates are found later in this section.

'Preliminary' fertilising

Organic manure is frequently laid on the pond bottom before flooding. Although some nutrients can be lost by this method it is considered to be suitable because it saves labour and can be done on a very large scale using heavy machinery. If this method cannot be used then the manure must be distributed from a boat. Boats specially designed for feeding, which have self-emptying hulls, are most suitable for this form of application of both organic and inorganic fertilisers. To save labour the manure should be loaded directly into the boat or piled by the landing jetty. When manures are required this way then the required amounts of inorganic fertilisers should be added to the load and applied at the same time.

> *Application rates for 'preliminary' fertilising:*
> Nitrogen (N) – ammonium nitrate or urea (carbamide), or in alkaline water use ammonium sulphate. These are added at 100–150 kg/ha.
> Phosphate (P) – superphosphate added at 100 kg/ha.
> Organic fertilisers (manures) – added at 2–3 tonnes/ha.

'Maintenance' fertilising

In the ideal situation where preliminary doses of fertilisers are distributed to the dry pond, maintaining fertilisers should be added every second week from a boat. Organic and inorganic fertilisers are mixed and applied together to economise on labour. If dry manures are used as the maintenance fertilisers then the breakdown and utilisation processes of this material in the pond are quite slow. In contrast, 'liquid' manures can be used. They dissolve quickly, and their effect appears faster and more effective (Fig. 5.3). The effect of preliminary ferti-lising is to produce large numbers of small invertebrate organisms which are continuously grazed by the fish. Since these food organisms can consume all of the available nutrients in a short period of time, their multiplication rate will quickly decrease. To prevent this from happening, maintenance fertilisation should be done as frequently as possible during the whole growing period, but at least every second week. Despite increasing consumption by the fish, it should be possible to provide an optimal nutrient level for food organisms, enabling them to multiply continuously, thus

Fig. 5.3   Fertilising with liquid manures (slurry) produces a rapid effect. This should be performed frequently with small doses.

providing a natural source of nutrition for the fish for a large part of the growing season. It is the role of the fish farmer to maintain the reproduction of plankton over the maximum length of time in order to keep up the level of natural food available for the fish. If the production is based on mono-culture (i.e. only common carp are stocked in the pond) the types of food organisms consumed are restricted mainly to the zooplankton, worms and insect larvae. This may result in the phytoplankton population multiplying fast. In extreme con-ditions the water turns green due to overproduction of the algae (an algal 'bloom'). These microscopic phytoplankton organisms produce oxygen during daylight hours but con-sume oxygen at night. As a result, over a 24-hour period oxygen levels are subjected to tremendous fluctuations, with the possibility of the minimum levels (found at dawn) falling to critical levels. In such instances fish mortalities can occur due to a lack of oxygen. The danger of algal bloom can be decreased by the activity of herbivorous fish which filter feed on the phytoplankton. As they graze on the phytoplankton population the balance between consumer and food organ-isms stabilises. Such unfavourable processes may also be stimulated by excess fertilising.

Application rates for 'maintenance' fertilising:
Artificial fertilisers – 20–30 kg/ha.
Organic fertilisers (in liquid form where possible) – 100–200 kg/ha.

Significantly higher dosages of fertiliser should never be added to ponds. If organic fertilisers are added at significantly higher levels, the breakdown processes which normally occur in the bottom mud may become anaerobic. This frequently occurs as a result of falling atmospheric pressure, which reduces dissolved oxygen levels in the water and can result in methane, ammonia and hydrogen sulphide being released. These toxic chemicals coupled with low oxygen levels can result in fish mortalities.

'Green' fertilisers are slower acting but cheaper forms of organic fertiliser. In the simplest forms, green fertilising can be achieved using grass cuttings, along with other bank side vegetation. It is particularly good for fry rearing ponds. Small amounts of plant fibres floating on the pond surface slowly decompose and provide an ideal environment for insect larvae, particularly mosquito larvae, and bloodworms. These all provide good food for fry. During their decomposition considerable quantities of nutrients are released slowly and evenly into the pond water, thus providing a continuous supply of nutrients to the environment.

The importance of organic fertilising

Organic fertilisers not only provide algae with their basic nutrients, but also directly provide a good source for invertebrate animals. The release of nutrients from the organic fertilisers is a relatively slow process, as the complex organic molecules are decomposed into simple molecules as a result of bacterial activity. However, practice has shown that organic fertilisers can affect the environment faster than these processes should allow (i.e. before decomposition is complete). The reasons for this are that the bacteria and small particles of organic material themselves provide food for small crustaceans and insect larvae, as they can filter these particles directly. So, in addition to promoting the growth of phytoplankton, there is a direct effect on the food supply for invertebrates. This effect is particularly apparent when using liquid fertilisers, such as the slurry from cows or pigs or solid manure steeped in a container of water. This can be used at say $5–10\,m^3/ha$ every two weeks.

5.3   Stocking and transportation

If the pond has been treated and prepared correctly, it should be rich in basic nutrients. Biological processes should have begun and the chemical content of the water should be stable. Stocking can now be considered. Fish are overwintered in ponds according to species, size or age group. As the temperature starts to rise in the spring, the fish become more active, their metabolic processes speed up and they start to feed.

The exact time to stock fish depends very much on the weather but it usually takes place in the middle of March when water temperatures reach 7–8°C. Fish should be stocked as soon as possible as each day without the fish feeding results in a loss in weight. This weight loss is not the only reason to stock fish early. During the winter the fish are generally weak. Their metabolism, although active, still leaves the fish susceptible to infection as their defence mechanisms are not functioning. (N.B. This is less of a problem in a pond situation, where conditions can be controlled, than with wild fish stocks.)

Wintering ponds are very suitable for the administration of quick 'bath' treatments. During the winter the farmer will regularly sample fish stocks for a health check. If parasites are found in the stock the overwintering ponds are suitable for treating with the appropriate chemicals for short durations, i.e. 'bath' treatments. The choice of treatment chemical/s and the concentration should be determined by the veterinary expert and administered with considerable care.

Fish harvested from the wintering ponds are counted or weighed into tanks and transported to the ongrowing ponds. Great care should be taken to ensure that tanks are not overstocked and that there is sufficient oxygen (see Table 5.1). Poor husbandry accounting for fish losses during transportation at this time is regarded as disastrous in stocks already overwintered, as their value is high following the overwintering and replacement supplies may not always be available.

It is particularly important to ensure that younger age group fish are not subjected to temperature shocks. The temperature of the transport tank water and the pond should be equalized by pumping or bucketing in pond water. If the temperature difference is more than a few degrees Celsius then this equalization will be necessary. Large differences may be found between the wintering pond and the receiving pond. Although this temperature balancing may take considerable time, it is nevertheless very necessary. Some species (e.g. the zander) are so sensitive to temperature shocks at this time of the season, that a 2–3°C temperature imbalance could cause significant losses.

The temperature-balanced water is released with the fish from the tank into the pond. Care should be taken to ensure that fish do not make contact with sharp edges or fall into the pond from a great height. It is best to use fibreglass chutes (Fig. 5.4). The farmer at the ongrowing establishment receiving the fish should examine them carefully. He should be sure that the fish are in good health and are the same sizes, weights and species etc. that were ordered. Any abnormalities should

Table 5.1 Data for fish transportation (weight of fish in kilograms per cubic metre of water with oxygenation and a transport temperature of 4–15°C).

Age/size	Species	Time of transport	
		2–6 hours	6–12 hours
		Weight of fish safely transported (kg)	
One-summer fish of 20–30 g weight	Carp	120	80
	Grass carp	130	90
	Silver carp	50	30
	Bighead carp	130	90
	Wels	140	100
	Zander (pike perch)	40	25
	Tench	70	50
Two-summer fish of 200–300 g weight	Carp	300	200
	Grass carp	325	225
	Silver carp	125	75
	Bighead carp	325	225
	Wels	350	250
	Zander (pike perch)	100	60
	Tench	175	125
Market fish of 1000–1500 g weight	Carp	600	400
	Grass carp	650	450
	Silver carp	250	150
	Bighead carp	650	450
	Wels	700	500
	Zander	200	120
	Tench	350	250

At 20°C or more weights of fish should be halved.

be noted on the delivery note and should be immediately reported. After the fish are released the pond should be examined carefully for losses. Dead fish sink into the mud and may only float to the surface several days later. If post-release losses are high, it should be reported to the farm manager who received the fish and an investigation should be started to identify the cause.

5.4 Supplementary feeding

In fish ponds, fish are kept in much higher densities than in the natural environment. As a result the natural food supply

Fig. 5.4   Fish are released into the pond from transport tanks using plastic chutes.

of the pond is not capable of providing adequate nutrition for the weight of fish. Additional (supplementary) feeds therefore have to be added to the pond by the farmer to compensate for this overpopulation (Fig. 5.5).

In the case of carp farming, many varieties of cereal grains are suitable as a source of supplementary feed. Carp are omnivorous according to their feeding habits, but they prefer whenever possible to consume insect larvae or other aquatic invertebrates.

In pond conditions feeds are added according to the size and age group of fish, starting with ground cereals, whole grains and eventually fermented cereals. As with all livestock, fish should be fed with good quality feed. It is a misconception that rotten, mouldy or rancid feeds can be suitable for fish. It is a false economy to feed poor quality foodstuffs to carp. In extreme cases fish stocks could be poisoned by contaminated food. At the very least growth rates would be reduced and the credibility of the fish farming industry would be questioned, particularly if the meat was tainted.

Feeds that are of reduced value to conventional agriculture, however, could be considered. The limiting factor should always be the nutritional value. Great care should be taken to ensure that the seed has not been dressed or treated with pesticides or other chemicals. Fish could concentrate such residues in their body tissues and become unmarketable.

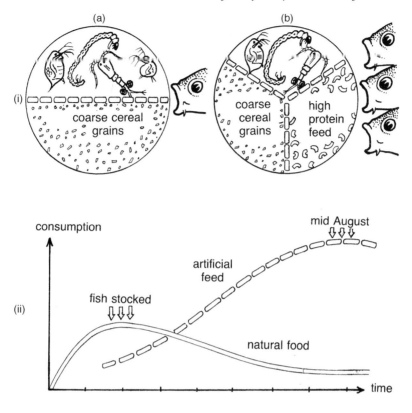

Fig. 5.5 (i) Supplementary feed requirements for (a) 'extensive' and (b) 'intensive' fish production. (ii) The amount of artificial feed is increased during the growing season.

Feed rates

The appetite of fish varies in response to many factors and consequently the daily ration of feed changes, according to temperatures, throughout the growing season. Appetite will obviously depend on temperature, but other factors are also important (Fig. 5.6). The carp starts feeding actively at 8–10°C. At lower temperatures its appetite is very limited; it eats little and infrequently. As temperatures rise, so does the appetite, reaching an optimum between 20–25°C. At such temperatures it can consume 2–10 per cent of its body weight per day of artificial feed (according to size/age).

C1 (one-summer-old carp): feed at a maximum of 10 per cent b.w./day

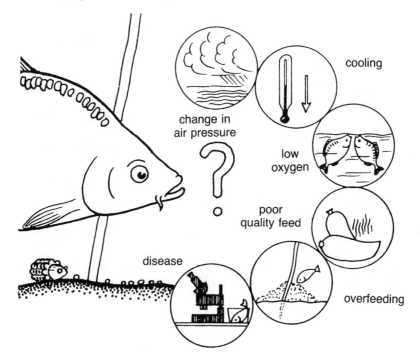

change in
air pressure

cooling

low
oxygen

poor
quality feed

disease

overfeeding

Fig. 5.6 A lack of appetite in a stock of fish may have a number of causes and should be investigated.

    C2 (two-summer-old carp): feed at a maximum of 5 per cent
      b.w./day
    C3 (three-summer-old carp): feed at a maximum of 2 per
      cent b.w./day

Supplementary feeding should start immediately after spring stocking, starting with low rates, and is distributed daily from a boat to fixed feeding stations identified by wooden poles (Figs 5.7, 5.8). Where feeding is performed manually, feeding poles are used close to the pond edge at the rate of five to seven per hectare. The observation and regulation of feeding should be performed three to four hours after the feed has been added. If the feed has all been consumed then the ration has been too small. If there is still a considerable amount of feed present then the residue should be examined again eight to ten hours later. If feed still remains then the ration was too high for the appetite of the fish and the ration must be reduced. This assessment should be performed throughout the growing season with a view to avoiding under- or over-feeding (Fig. 5.9). If this assessment is

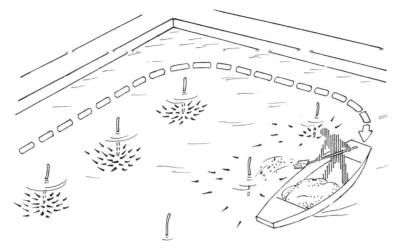

Fig. 5.7 Manual feeding is performed by distributing the feed to 'feeding stations' (marked with wooden poles).

Fig. 5.8 Open bottom boats can be used to produce a feeding 'strip'.

neglected or undertaken with insufficient care then either fish yields will be less than optimum or, in the case of overfeeding, costly feed will be wasted and the production costs per kilogramme of fish will rise.

Cereal feeds should be soaked prior to feeding. This should be done on the feeding boats where the daily ration is prepared the previous day and mixed with pond water. During the night the seed grains absorb the moisture and become soft.

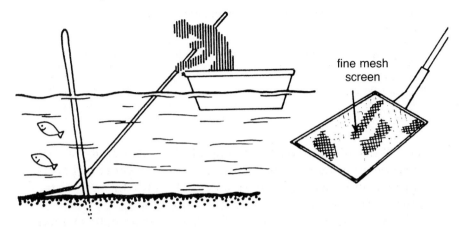

Fig. 5.9 The consumption of feed should be checked regularly. Daily feed rates should be increased or decreased accordingly.

Fish prefer cereal in this form and can crush and digest it more efficiently. With large older fish this stage could be omitted if it proves difficult to organise.

During the planning of the farm management, it is important to avoid unnecessary movements of grain. If possible large quantities should be stored in silos from which the grain can be released into the feeding boats.

In the spring it is sometimes necessary to feed medicinal feeds (e.g. reinforced with vitamins) for short periods. Medicinal additives are used to combat various disease problems recognised by the veterinary surgeon. Vitamin supplements in the spring are frequently used to improve the condition of fish that have overwintered poorly. This may be particularly valuable in situations where infectious diseases occur.

The ration size

During the season fish grow evenly in the pond. The daily feed ration should be increased to keep pace with this growth (see Table 5.2). The growth potential of the fish can be optimised by avoiding underfeeding. If fish are feeding actively, feed rates should be adjusted quickly to meet requirements. It is particularly important when there is a large amount of natural food produced by effective fertilising. Under these circumstances fish can confidently be fed with a high ration of

Table 5.2 Ratio of monthly supplementary feed as a percentage of the yearly total.

Age groups	March	April	May	June	July	August	Sept	Total
One-summer fish	–	–	2	3	15	40	40	(100%)
Two-summer fish	2	5	10	20	25	25	13	(100%)
Market fish	2	5	10	20	25	28	10	(100%)

feed. If fish are underfed, when plankton is abundant, the fish will consume more of the valuable plankton protein than is optimal, and it would be used not only for growth, but also for energy. This would be an inefficient use of feed.

Despite the best efforts of the fish farmer, natural food resources noticeably decrease by August. When this natural food component of the diet is reduced, the farmer has to consider the problems of overfeeding cereals which are used as compensation. If fish feed on too high a proportion of cereal food, there are several dangers. Most seriously, fish become too fatty because of the excess starch in the diet. Although this is not a problem for younger fish that can utilise fat as an energy reserve during the winter, marketable size fish which are too fatty are regarded as of inferior quality and their market value is affected. In husbandry there are also disadvantages. Without the necessary proteins available the overall feed utilisation becomes inefficient. Although the fish continues to eat large quantities of feed, the food conversion ratio (FCR) is poor.

Fish health can also be affected by this unbalanced diet. Physiologically, the fish slowly lose their ability to resist disease and become susceptible to bacterial and parasitic infections. This is particularly noticeable with fingerlings at the end of the first season.

When protozoan parasite levels build up in the fish population, this further affects the fishes' ability to utilise feed efficiently and the overwintering stock can be weakened further. The nutritional status of herbivorous fish can also be affected in multispecies ponds (polyculture).

The result of overfeeding cereals to common carp also applies to the herbivorous grass carp. If they do not receive enough green vegetation at the end of the summer, grass carp suffer inflammation of the intestine (enteritis) due to the consumption of cereals. This causes significant losses. To avoid this problem and to produce a healthy grass carp population at the autumn harvest, grass carp should be stocked according to the plant population of the pond. In the case of overstocked populations, additional green vegetation should be fed at least during the second part of the season.

The estimation of feed consumption for bighead carp is difficult to determine and influence. Bigheads visit the feeding places of the carp and whilst filtering algae and zooplankton from the pond they also take up small particles of 'flours' or cereals which are inevitably mixed in with the prepared cereal feeds.

5.5   Yield estimation and sample fishing

Fish farmers have more difficulties than conventional farmers in estimating the size and weight of their stock at any given time, and in predicting the total yield at the end of the season.

The substrate that fish live in is water, which is an alien environment for humans who, consequently, have to rely on secondary signs to provide information about what is happening beneath the surface. Stock numbers cannot be counted exactly; they can only be estimated. Although such rough estimations can produce significant mistakes, they still have value since production control must be based on the best available information at any particular time.

Estimation of yield is based on continuity. This means that fish farmers have to determine production from extrapolating data recorded from when the fish were stocked.

Data such as loss estimation after stocking should be continuously recorded. Information such as feeding rate, bird predation (i.e. presence of gulls on the pond) should also be noted. It is also important to assess the development of populations of natural food organisms. Such observations are used by experienced farmers to make theoretical assessments of stock. However, such estimations should be strengthened by practical means such as sample fishing.

Yield estimation

At sampling, the fish farmer retains a sample (of, say, 100 fish) from the population for analysis. The average weight of fish from the sample is determined and using the stocking figure the current stock weight per hectare can be calculated. This weight should then be compared to that previously obtained and the weight gain can be determined. By dividing the weight gain by the number of days, the daily weight gain can be calculated.

According to the amount of feed supplied, the farmer can also calculate the feed required to produce one kilogramme of fishmeat.

To evaluate production efficiency in a fingerling pond over a two-week period

At the start of this period, the average weight of the carp has already been found to be one gramme by means of previous sampling. As the stock of advanced fry was 100 000, the standing crop is 1 g × 100 000 = 100 kg.

At the end of the period, two weeks later, it can be determined from sampling that the carp now average 2 g. The standing crop is now 2 g × 100 000 = 200 kg, and the production during the last two weeks has therefore been 200 kg − 100 kg = 100 kg.

An average daily ration equal to 15 per cent of the fish biomass present at the beginning of the period has been fed, and 0.15 × 100 kg = 15 kg of feed has been distributed each day.

Therefore, during the entire period under examination, 15 kg × 14 days = 210 kg has been fed. This 210 kg of food has resulted in the production of 100 kg of fish flesh, and therefore the food conversion rate is 210 kg:100 kg = 2.1:1. This represents efficient production during this period.

Stock sampling

Stock samples can be collected in different ways. The simplest method is to use a throwing net (Fig. 5.10a). The mesh size of the throwing net depends on the age and size of the fish. Several hundred fish should be caught from the pond two to three hours after feeding, and the sample measured and analysed on the boat. The feeding areas should be approached cautiously since the fish will detect noises and vibrations which travel quickly and over great distances through water, and they will simply swim away.

Occasional weather changes, strong winds, etc. prevent an adequate sample from being caught and only a poor estimation can be made. Data produced from such estimations can be dangerously misleading.

Once a month, seine nets should be used in addition to the throwing net. It should be used to catch a sample of 1000 fish around the feeding station (Fig. 5.10b).

(a)

(b)

Fig. 5.10   Sampling for (a) small numbers and (b) large numbers of fish.

During the sampling process, it should be noted that fish of different sizes and weights may colonise different parts of the pond. Fish grow better in deeper parts of the pond and in areas where feeding takes place. Fish taken from such areas could thus produce an unrepresentative sample of the pond population.

Care should therefore be taken to ensure that the sample should reflect the growth characteristics of the entire pond. It is convenient to catch fish from feeding places since they collect there in large numbers (making it convenient for sampling). Those fish living in other parts of the pond may be spread too sparsely to make capture practicable. Fish caught from the feeding station will usually have an intestine full of food. The weight of this food should be discounted from the measured body weight. If this is not taken into consideration the weight of the stock could be overestimated by as much as 15–20 per cent (depending on size and age group).

Catching common carp for sample weighing is a straight-forward process as the fish can be caught easily. This is not the case for the herbivorous carp species (grass carp, etc.). Herbivorous fish are extremely fast and difficult to catch during the summer in warm water. Problems are increased by the fact that the fish will only stay near feeding stations if they are short of natural food.

Consequently only small numbers of these fish may be caught by seine netting or using throwing nets, and the estimate of weight gain and population size is less accurate.

Sample fishing is a very important aspect of management and should be performed every two weeks. It can also be used to provide samples for health monitoring.

5.6   Oxygen supply

On farms where production levels are very high, ponds cannot support stock without additional aeration. It is a characteristic of pond culture that additional oxygen is required, not for the entire season, but for certain periods, and certain times of the day.

As a normal feature of fish production on the farm, ponds are understocked in the spring, as the size and weight of the fish are small and their metabolism slow in the cold water. In the middle of the season, fish fill the available space to an optimal level. By the end of the season, fish have continued to grow and consequently the ponds are overstocked (Fig. 5.11). For example:

If a pond is stocked with 5000 fingerling fish of 40 g body weight in a one-hectare pond (surface area), with an 80 per cent survival rate, and the fish reach 200 g body weight in

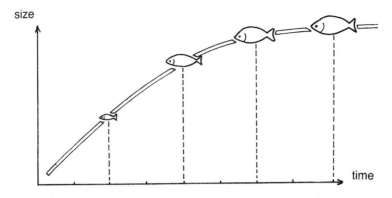

Fig. 5.11   The initial rapid growth rate slows down as the natural food supply diminishes, and water cools down in the autumn.

August, there will be 800 kg (4000 × 200 g) biomass per hectare by then.

If these 200 g (C2) fish are then stocked the following spring for their third season's growth, only 1000–1500 fish should be stocked per hectare to allow them to reach their required one kilo market weight as soon as possible.

By virtue of this, at the start of the season there is a much lower biomass of fish than at the end of the season, since the number of fish must be balanced by the weight of the fish.

Because dissolved oxygen concentrations are unstable the critical period is always the second part of the season, when the pond becomes 'overstocked'. It is worth noting that in very intensive conditions fish stocks can be greatly overstocked, providing there is a through-flow of water and oxygen is supplied continuously. In such systems space for growing is helped by the frequent harvesting.

Oxygen balance

During pond husbandry, oxgen availability is governed by not only the fish stocks but also the daily requirements of natural organisms within the pond. Oxygen is utilised predominantly by populations of algae and bacteria, as well as fish. The more intensive the system of production (i.e. a greater biomass of fish requiring greater weights of supplementary feed) the greater the influence these organisms have upon the oxygen levels. It is well known that aquatic vegetation (algae and submerged plants) produces oxygen as a by-product of photosynthesis in daylight hours, but like all aquatic organisms will consume oxygen at night. The amount of oxygen in the water is therefore influenced not only by fish but also by all other organisms living in the pond (Fig. 5.12).

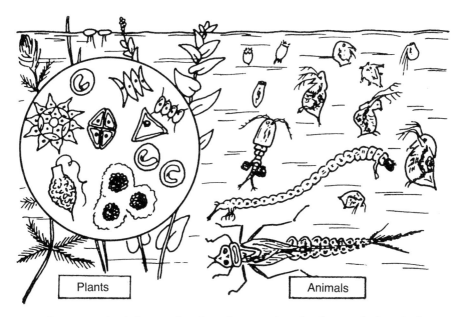

Fig. 5.12 In fish ponds, the plant and animal population is in equilibrium. Their continued growth and development are the basis for economical fish production.

The development of dangerous oxygen levels cannot be forecast accurately. Conditions that favour oxygen deficits do, however, develop mostly in the second part of the season (Fig. 5.13). This is because water temperatures are high, the weight of the fish in the pond is at its maximum level, and thus the input of feed into the pond is at its highest. The first duty of pond fish farmers is to prevent such situations arising and to avoid fish mortalities, rather than to provide oxygen on a continuous basis as this may not always be cost-effective. Equipment and solutions should be chosen accordingly.

In intensively farmed ponds, it is advisable to use a meter to measure dissolved oxygen levels on a daily basis. Potentially dangerous situations can then be predicted and dealt with immediately.

Fig. 5.13 When the pond is coloured very green as a result of an algae bloom, fish may be under severe stress at dawn due to very low oxygen levels.

There are several simple techniques used to counteract dangerous oxygen depletion. In small ponds, the simplest method is to start or increase the flow-through of fresh water which itself can be aerated by spraying it into the pond. Also, in small ponds, other aeration devices that can pump and spray the water can be used. Because such pumps only have a small effective radius, they would not be suitable for large ponds. When oxygen depletion is expected in August (when there is a high stocking density, high temperatures or intensive algal growth) or where the local conditions cause regular annual difficulties (e.g. muddy ponds or water supply problems) it is advisable to install specialised oxygenation equipment which can be used when required.

For use against seasonal depletion, specialised aeration equipment is very effective. Several systems are available, which include:

- spraying water (using paddlewheels or sprayers) (Fig. 5.14);
- injecting air into the water (using a surface-mounted pump combined with a Venturi system: Fig. 5.15a,b);
- using fans or blowers to push high volumes of air at low pressure through specialised aerators.

All of them can be very effective in shallow ponds as the pressure of the water column is small and energy requirements are kept to a minimum. They are usually run through

Fig. 5.14 Paddlewheel aerator used to spray water into the air. An efficient aerator in a pond, but increases water loss due to increased evaporation.

an electric timer so that they switch on at night and switch off early in the morning, when they are particularly useful at preventing 'dawn depletion'. Each system has merits and disadvantages, and thus the choice will depend upon personal preference and availability.

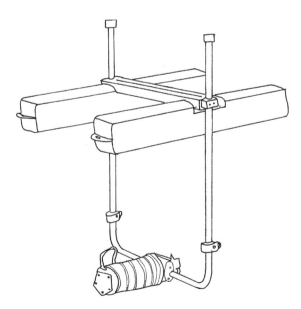

Fig. 5.15(a) Floating air injection aerator. The depth and angle of the pump can be adjusted.

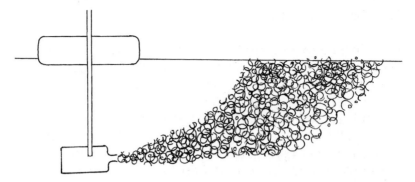

Fig. 5.15(b)   Efficient aeration produced with the Venturi system.

Long pipe runs can also be used to distribute air to the furthest parts of the pond to ensure that aeration reaches all areas (see Fig. 5.16).

In places without electrical power, diesel engines may be installed close to ponds to drive compressors or other electrical equipment.

Fig. 5.16   Irrigation systems may be used to aerate water close to the edge of the pond.

5.7 Health control

In the pond environment, fish are much more densely populated and thus more susceptible to different diseases such as parasitic and bacterial infections. This situation is exaggerated by the feeding of artificial food.

Artificial feeds (particularly cereals) are not a natural food for carp. Fish fed on cereals (which have a high proportion of starch) can develop various ailments associated with this diet. In its mildest form fish may deposit fat and become weak and more susceptible to infections. In more extreme cases inflation of the intestine and liver damage may occur. Considerable care should therefore be taken to observe stock in intensively farmed ponds as disease problems can develop quickly and more readily.

Primary signs of disease

Fish farmers should frequently observe stock and look for unusual behaviour in the ponds.

The most sensitive organ of the fish in terms of abnormalities or infections is the gill. If a disease or water quality problem manifests itself on the gill, gaseous exchange is impaired and fish instinctively search for layers of water more abundant in oxygen where they can breathe more easily. In such cases fish will gather at inlets or near the surface of the water. There they will stay for long periods, 'gasping' on the surface. If such abnormal behaviour is detected then a sample of fish should be taken from the inflow area and taken for veterinary investigation. As the reactions of sick fish are impaired, they can be caught much more easily than healthy fish. They can easily be caught with a throwing net or long-handled net.

Secondary signs of disease

As part of the management of big ponds (i.e. those with very large surface areas) fish farmers should be aware of secondary signs regarding the state of the fish population. For example, the behaviour of gulls, particularly on fry rearing ponds, can provide information regarding the health of the fish. If gulls appear in large flocks on ponds or parts of a pond, their behaviour should be closely observed. Gulls can only catch healthy fry in shallow water (for instance, during autumn harvests). Usually they concentrate on sick fish that are swimming slowly near the surface. When gulls are circling above the pond, but are not attacking fish, the fish are staying below the surface layers. The fish are probably searching for the oxygen-rich upper layers of the pond, but they are still quick and healthy enough to recognise the shadow of a diving bird and escape. This could mean the early stages of a parasite infection which has not yet become widespread. Early detection and treatment by the veterinary surgeon should easily cure the problem at this stage.

When gulls are observed swooping down into and under the water then the state of the fry is weaker and the infection is more advanced, since the fish are unable to escape the gull attacks (Fig. 5.17). If an infection has become as advanced as this then significant losses will be the result.

Fig. 5.17 The behaviour of gulls can give an indication of the presence of parasites in the fish stocks.

Sampling

Samples of fish which look sick or are suspected of having a disease should be sent immediately for examination by a fish disease specialist. Only live fish are suitable for veterinary examination. During a long period of transportation weak fish can easily die (particularly if not packed properly) making them unsuitable for examination. In such cases a further sample will need to be collected and transported which takes time and delays treatment still further. Fish should therefore be carefully packed in a plastic bag half full with water and with oxygen filling the other half. Ice can help to cool water during hot periods. At the time of sampling, details of symptoms and behaviour should be reported and the data presented with the sample for investigation.

Treatment of fish

When results of the veterinary investigation are known, the fish farmer has to prepare any chemicals recommended for treatment.

One of the greatest disadvantages of pond husbandry (particularly in the case of big ponds) is that it can be quite difficult to prevent or treat disease problems successfully. Effective treatment possibilities are limited. A change of water may solve some simple problems, for example, when disease is a result of poor water quality in the pond. However, large quantities of additional freshwater are not usually available for the majority of farms. Bacterial diseases can be readily controlled using medically treated feed. Medicated feed has to be prepared according to veterinary prescription in terms of the medication, the dosage and the rate of feeding. After or during the treatment, samples must be analysed to determine if the treatment has had the desired effect.

Diseases can be prevented (particularly when the levels of ectoparasites build up during the summer period) by the use of certain chemical treatments which are themselves quite poisonous to the fish. These are also poisonous to man and should only be measured or handled by experienced personnel.

Apart from the simple chemical treatments described in Appendices 8 and 9, the diagnosis and treatment of fish diseases are, however, not covered in any detail in this book. Please refer to specialist texts.

5.8   Autumn harvesting and sorting

During the autumn, the cooling of the water temperature will gradually cause the fish to cease feeding. They migrate to deeper parts of the pond and prepare for winter. In Central Europe water temperatures are usually low enough to start harvesting at the beginning of October.

Before harvesting is started there is much preparatory work to undertake. Equipment such as ropes, nets and sorting tables has to be serviced. Seine nets and dip nets should be repaired. Particular care should be taken with protective clothing. Boots and waders, waterproof clothing and gloves should be checked, repaired and/or replaced. A fish farmer cannot work effectively with faulty clothes or equipment and efficiency will be lost. Harvesting schedules should be prepared for the work teams under team leaders. The harvesting of particular age groups of fish is planned according to the demands of the market. The wintering ponds are prepared for the reception of harvested stock. Ponds that were kept dry in summer are cleared of all vegetation which is either burnt or removed. If wintering ponds were kept flooded (e.g. because they were used for fry rearing) they are drained and sterilised with quicklime. Dams, inlets and outlets are checked and damage is repaired.

Before flooding, the wintering ponds should all be sterilised with strong disinfectants such as quicklime, or better still, 'bleaching powder' $(CaCl(OCl).4H_2O)$ (also known as chloride of lime). Pumps and compressors used during the winter period should be mended. After completing these works, the harvesting teams are ready for the autumn harvest and the wintering ponds are ready to receive the fish.

Draining

The harvesting begins with the draining of ponds. Draining should be done through a screen of a size determined by the size of the fish in the pond. These screens must be cleaned continuously during the draining process. In the second phase of draining, to avoid fish mortalities in the shallow water, a 24-hour watch must be organised. Bird scarers should also be on hand for ponds with very large surface areas.

In the summer, gulls can only endanger sick fish, but in autumn fingerlings are unprotected against such birds in the cold shallow water. Flocks of gulls can take large numbers of fish in the early morning during draining. Although they cannot catch the larger fish, they can still cause a great deal of physical damage to them (Fig. 5.18). Sadly, birds cease to be frightened by bird scarers after a certain period of time. When this is the case a nightly guard should be employed to frighten birds away. Such a guard also provides security against poachers, as fish concentrated in small areas are easily taken.

Fig. 5.18   Birds may cause considerable damage to slow moving fish and should be deterred.

During draining fish are drawn into the outer or inner channels. The fish farmers then harvest fish from the channels using a seine net. Harvesting channels gradually fill with mud over the years and the pulling of seine nets in these channels becomes physicallly very hard work. One half of the team works the bank side (near the reed beds) while the other team works along the 'muddy' bank in the pond (Fig. 5.19).

The latter group has to work much harder as they are working in soft mud. The net may frequently cut deeply into the mud and the mesh blocks. In such cases a boat follows the

Fig. 5.19 Fishermen pulling a seine net through the harvesting channel. A boat follows the float line in case the net becomes snagged.

net in. When it becomes stuck it can be pulled upwards to free it. Occasionally, a net may need to be pulled for several kilometres. Workers may frequently need to rest!

In ponds formed from flooded valleys, the original river bed will act as the harvesting channel. This also fills with mud during the year and working in the soft pond mud becomes very difficult.

Recent developments in harvesting technology have led to the design of external harvesting 'units' servicing several ponds. Because of the cost-effectiveness of such a system it is important to build it to a high specification (e.g. concrete floors, covered sorting areas, etc.). Ponds and ramps should also be built to allow suitable access to transport lorries, etc. This represents a particular advance over the use of sodden, muddy roads during the wet autumn period. In addition, such harvesting devices usually have the potential to deliver good quality fresh water supplies to the fish stocks during the drainage and harvest process.

At the last stage of drainage a seine net is used to force fish

Fig. 5.20 Fish captured within the pond (rather than in a fixed device behind the dam) require portable equipment to ensure that conditions are effective for harvesting.

from the harvesting channel, into the device where it is pulled into a corner (Fig. 5.20). The float line is suspended on sticks after the fish have been crowded into the edge. Harvesting can then begin.

Sorting and measuring

Fish are collected in large bins (40–50 litres) which are carried by two workers to the sorting table where they are separated into species and by size (Fig. 5.21). If there is a particular size or species which forms the majority of the catch then other species are removed before filling the bucket. The full bucket is then weighed and emptied immediately into the transporter.

In ponds that contain several species of fish (polyculture) a major sorting job is inevitable. This is the only dis-advantage with this system compared with monocultural

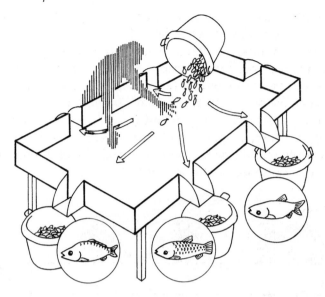

Fig. 5.21   A sorting table for separating different fish species.

carp production. The only time when sorting would not be necessary is when the mixed harvest is to be restocked at the same stocking ratio into another pond. This can only happen when previous production rates fit into the next season's plans.

Sorting is a very important aspect of fish production. Fish taken onto the sorting table are separated by workers who are each responsible for sorting one size of species. Separated fish are collected in large bins.

The measurement and weight of fish stocks are an integral aspect of the harvest. Once a bin has been filled fish are transferred into another container (in a basket) which allows water to drain away. The basket is weighed and the fish transferred to the transport vehicle and subsequently taken to the overwintering or marketing site. Wintering may be done in production ponds (in the case of autumn stocking) or in specially designed ponds.

In the case of more sensitive or fragile fish such as fingerlings, fish should only be weighed in water-filled containers, the weight of water having been recorded before adding the fish. A simple calculation allows the weight of fish to be determined. Fish can still be damaged in cold water in spite of their reduced metabolic activity and care should be taken to ensure that they are not overcrowded and allowed to deoxygenate.

As with other fish farming practices, much hard physical labour can be replaced with the aid of mechanisation. Many types of fish lifts, conveyor belts, automatic graders and weighing machines can be used during sorting and weighing. Sorting can be made more efficient with the use of an endless conveyor belt on which the fish are moved along. Fish pass between sorters standing on either side of the belt, each picking out their respective size or species. The most abundant variety should be left on to drop off the end into a large collection container or storage tank (Fig. 5.22).

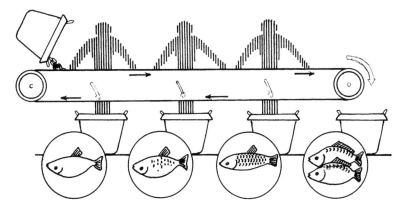

Fig. 5.22 Sorting can be automated on large farms by using conveyor belts.

Health control at harvesting

Harvesting provides a good opportunity for health-checking fish stocks. If necessary short bath treatments can be used to eliminate ectoparasites. Such bath treatments are also useful as a prophylactic control for fungal infections that may develop as a result of the handling damage that occurs during harvest (see Appendix 8).

The optimum period for a short 'bath' treatment is during transport to the wintering ponds. Effective treatments can be provided using mixtures of various chemical treatments at this stage. If the transport is longer than a few minutes the mix should be administered only 5–10 minutes before the fish reach their destination.

The production process is finished after the sorting and weighing of the stock. The cycle continues with marketing and wintering of stocks to be reared further.

5.9   Wintering as a technological operation

Wintering is the final aspect of the autumn harvest. It can take several months and is an extremely important aspect of the farming cycle. During this period fish reared during the season are kept in high densities in relatively small ponds (600–1000 m^2). See Table 5.3 for stocking ratios.

Table 5.3   Stocking density in wintering ponds.

Age groups	Number of fish per m^2	Average weight kg/m^2	Water inflow per 100 kg fish (litres/minute)
One-summer fish	80–400	4–8	7–10
Two-summer fish	40–60	8–12	6–8
Market size fish	7–10	6–8	6–7

The fish do not require feeding as their metabolic rate is very low at winter temperatures. Oxygen requirements are still relatively high so overwintering ponds must be fed by a continuous through-flow of fresh water. The through-flow of water, as well as providing oxygen, also removes any build-up of metabolic waste products such as ammonia. No matter how slow the fish metabolism may be, the accumulation of waste products becomes significant because of the high density of stock. Aeration is therefore not a suitable alternative to an input of fresh water.

Wintering ponds have to be specially prepared for use. After cleaning, a comprehensive sterilisation should be performed. If this is done using lime, it must be thoroughly washed out before fish are stocked. It should then be treated with malachite green or formalin to prevent parasites from surviving in the margins of the pond (Fig. 5.23). After this fish are stocked.

During the autumn harvest fish are sorted into sizes and species. Similar stocks from different ponds will consequently be mixed together for overwintering. Fish from different origins may have varying resistance to parasitic diseases and cross-infection may occur.

Physical damage caused to the fish during the harvesting and sorting processes is susceptible to infection by fungi. Because of this, considerable importance should be placed on

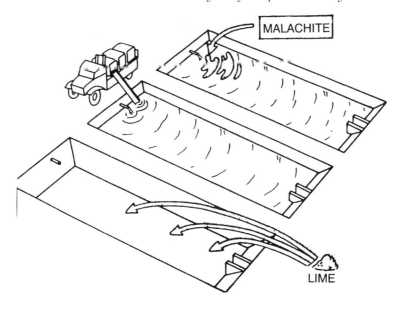

Fig. 5.23 Wintering ponds are prepared with lime prior to stocking fish. After ponds are stocked they are treated with malachite green.

the use of prophylactic treatments prior to stocking into the wintering ponds. The most effective treatment is to use malachite green once a week. The treatment concentration is used at 0.1 mg/litre (i.e. 100 g per 1000 m³).

The malachite green crystals should be carefully dissolved in water. If this is done outside, prevent the wind from blowing the crystals over personnel. Protective clothing should always be worn. The calculation, measuring, dissolving and application of the chemicals must only be done by suitably trained personnel.

Later during the season these treatments can be reduced to a fortnightly application but it is important that it is administered continuously throughout the whole wintering period.

The wintering period is a less intensive period of work for fish farm personnel. There should be time for the mending and making of nets. Occasionally batches of marketable size fish are removed for sale. Daily work before the ponds freeze over involves the prevention of bird damage and the removal of dead fish. All mortalities should be recorded in the farm log.

Fish are in high densities in wintering ponds but their activity is usually greatly reduced in the cold water. Young fish, however, rarely lie on the bottom in winter and may be easily caught by gulls. If fish are seen to be moving frequently in the pond it could indicate the onset of a disease problem. The reasons for the activity should be investigated and appropriate control measures should be taken.

During periods of extreme cold, ponds freeze over – often for long periods. Ice does not form around the inflow because of the constant movement of the water, but should be removed from around outlets, pipes, etc., to avoid damage. To avoid accidents to personnel during this period they should always work in pairs and strict safety procedures should be enforced (using safety ropes, etc.) (Fig. 5.24). Spreading plates may be fixed under the water inflow to increase oxygenation in both the wintering ponds (Fig. 5.25).

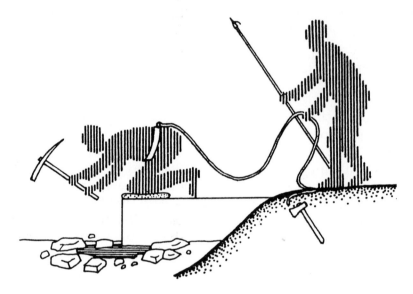

Fig. 5.24 After freezing, areas around the inlet and outlet devices should be de-iced daily.

Ice holes should also be made to provide, say, $50\,m^2$ of clear water per hectare of pond. If oxygen depletion occurs under the ice, insect larvae and frogs appear first (these overwinter in the bottom mud). Fish will be evident later gasping at the surface. Remedial action must be taken otherwise large stock losses will occur. Addition of further fresh water at this stage is obviously a solution when it is available, although aeration using compressed air is efficient. If

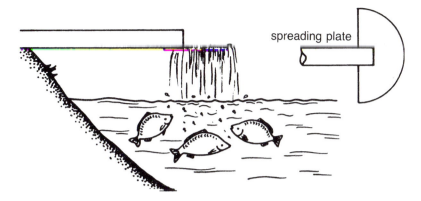

Fig. 5.25 Inlet supplies can help with aeration by the use of a simple spreading plate.

there is thick snow on the ice it prevents light penetration into the pond below. Oxygen production by algae then slows down or completely stops. When this happens only oxygen consumption occurs under the ice, with no production at all during this period. Care should also be taken to control the health of the stock during this time. Samples of fish appearing around the ice holes should be sent for checking. Although diseases should develop more slowly in the cooler water, they are also slower to cure, and regeneration also takes longer.

There is a risk of parasite loads increasing due to the high density of the fish stock. The cool environment also can be advantageous to certain bacterial and viral infections. Any signs of abnormality which could signal the development of a disease in the stock should be investigated with care. Abnormalities should be immediately reported to the farm manager.

During the spring when ice and snow begin to melt, fish start to become more active and begin to search for food. This can be detected by the increasing turbidity of the water and visible movements of the fish. At this time the overwintering period is regarded as over, and younger fish should be restocked into the ongrowing ponds.

The harvesting of the wintering ponds should be started when they are full and continued with the pond in a half-emptied state. An important aspect of restocking is the weighing of stocks and sorting according to size (Fig. 5.26). As fish have not been feeding during the winter there will be a

Fig. 5.26 After the overwintering period stocks should be harvested. Fish are sorted again if necessary into species or size and weighed into their ongrowing ponds.

small but acceptable loss in weight of the stock. The weight of fish prior to wintering minus their weight after wintering is known as 'wintering loss'.

Chapter 6
The Propagation of Other Pond Fish Species

6.1 The propagation and rearing of herbivorous fish

Feed utilisation of herbivorous fish

The three species of herbivorous carp referred to as the Chinese 'herbivorous' fish comprise the grass carp, silver carp and bighead carp. They originate from South East Asia and have now been naturalised into the fisheries and waters of Central Europe. They can feed on different food sources not utilised in a monoculture system using the common carp.

The grass carp consumes higher aquatic plants (macro-vegetation) that frequently invade fish ponds (Fig. 6.1). The silver carp filters unicellular algae produced in the ponds (producing valuable fish meat very economically!) (Fig. 6.2).

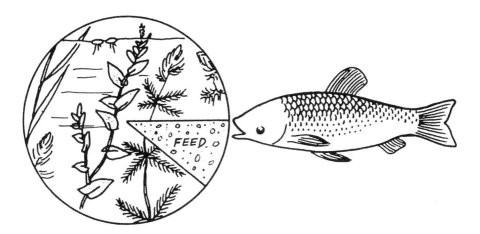

Fig. 6.1 The herbivorous grass carp feeds upon higher plants, as well as consuming some of the supplementary feed used for the common carp.

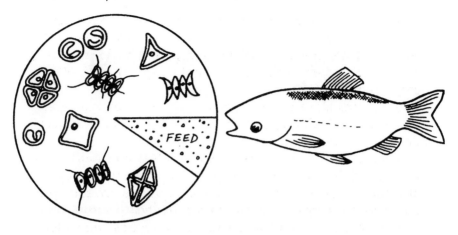

Fig. 6.2 Silver carp produce fish meat very economically in fish ponds as they only feed on algae (the lowest source of energy in the food chain). This species is only capable of filtering the smallest particles of artificial food.

The silver carp can be stocked at higher densities than the other species. The bighead carp filters larger algae (e.g. colonial forms), blue green algae as well as zooplankton such as rotifers and small crustaceans (Fig. 6.3). In consuming zooplankton, the bighead thus competes with the carp.

Such fish that naturally exist in flowing water are able to produce eggs in ponds but are unable to reproduce. In a

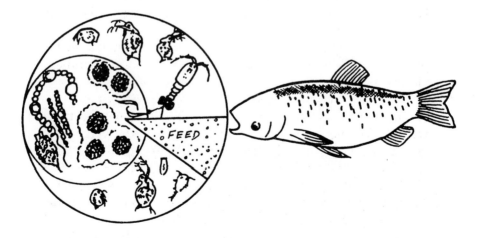

Fig. 6.3 The bighead carp is not exclusively herbivorous as it filters animal organisms from the plankton. It can also filter small particles of artificial food.

temperate European climate maturation requires five to eight years. Grass carp prefer nutrient-rich ponds overgrown with aquatic plants. For artificial feeding grass and legumes are suitable. The daily requirement can be as high as 15–20 per cent of body weight.

Silver carp like water of 'medium' fertility, coloured green or brown with algae. Such ponds require frequent applications of nitrogen fertilisers.

Bighead carp prefer small ponds with deep muddy bottoms. In addition to consuming planktonic organisms they can also filter organic particles. Organic fertilisers are thus preferred to the inorganic ones used for silver carp above.

Broodstock can be kept separately by species and so food demands can be simply provided. However, it is more common to keep the species mixed. In such cases there is usually an emphasis on one species, with others kept as supplementary species. It is also common to keep female common carp to 'cultivate' the pond bottom as they vigorously 'turn over' the top 10 centimetres of mud in search of food. In favourable conditions several hundred spawnable fish can be kept in one hectare.

Reproduction

The three species all show similarities in their reproduction. Herbivorous fish become mature for spawning in the early summer (early June).

Males are easily identified by the rough tubercles on the pectoral fins. The belly of the female is plump. Carp pituitary glands are used to promote the reproduction of the herbivorous fish. The injection technique is performed as with the common carp. The preliminary dose is 0.2–0.3 mg/kg body weight. To determine the decisive dose the body weight and girth also have to be calculated (Fig. 6.4). To inject very large specimens additional pituitary extract is required. Fish with a girth of 500 mm require 5 mg/kg of hormone, not the usual 4 mg/kg. Females of 600 mm girth require 5.5 mg/kg. Males like the common carp are also only injected once, with a dosage of 2–3 mg/kg. This is performed 12–24 hours prior to stripping. After treatment males and females should be kept separately.

The optimum temperature for ovulating in silver and grass

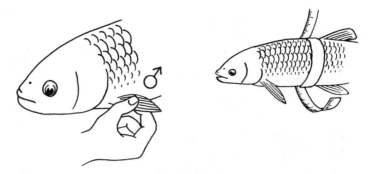

Fig. 6.4 The male pectoral fins of herbivorous fish should feel rough to the touch. Females have a rounded body. The girth is taken into account when calculating hormone dosage.

carp is 24°C, but with bigheads it should be 25°C. Egg maturation occurs in 210–220 degree hours (9–10 hours) after the decisive dose for grass and silver carp, but bighead carp require 240–250 degree hours (10–11 hours). [N.B. Degree hours ÷ temperature (°C) = hatching time.] During egg maturation the treated fish must not be disturbed. The appropriate oxygen level (>6 mg/l) is maintained by constant through-flow (at 4–6 l/min per female). Ripe eggs must be stripped without delay as they become over-ripe very quickly and will be unsuitable for fertilisation. The stripping of eggs should be performed under anaesthetic (using the same method as used for the common carp). Particular care must be taken when handling silver carp as they are easily over-anaesthetised and can easily deoxygenate in holding tanks.

Eggs are stripped out into a clean dry plastic bowl. The eggs must be run down the side of the bowl (not allowed to pour directly onto the bottom!). After this, milt is collected using a pipette or small glass tube. For every 1000 grammes of dry eggs, 2 × 5 ml of milt derived from a mixture of males is added and thoroughly mixed. Fertilisation occurs later when 100–150 mls of clean water are added and mixed.

Sperm cells are activated by the fresh water which maintains the fertilisation. In the water the eggs start to swell. Eggs from herbivorous fish float, swelling 50–60 times their volume in 1–$1\frac{1}{2}$ hours. Eggs are kept stirred until they swell to 2–3 mm in diameter, and then the eggs are rinsed several times and transferred into the Zuger jar (Fig. 6.5). Partially swollen eggs are added at the dry weight equivalent of 40–50 g/jar. The total dry egg weight should be measured so that the total

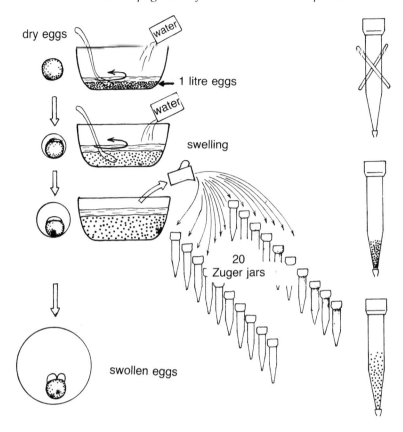

dry eggs

water

1 litre eggs

water

swelling

20
Zuger jars

swollen eggs

Fig. 6.5   Eggs of herbivorous species swell until they reach 2–3 mm in diameter. They are then stocked into Zuger jars where they continue to swell.

number of Zuger jars can be calculated to hold the entire egg mass.

The eggs of herbivorous fish are very sensitive to mechanical damage in the early development stages (for the first 10–12 hours following fertilisation). For the first 8–10 hours the water supply of the incubating jars should be adjusted to 0.2–0.3 l/min (measured by collecting the outflowing water in a measuring vessel). After the early phases, the water flow should be increased to 0.7–0.8 l/min to meet the higher oxygen demand of the eggs. During this second period the cellular structure of the unfertilised eggs will break down and the specific gravity will change, allowing the dead eggs to float above the living ones. It is usual for the eggs of herbivorous fish to have a lower fertilisation rate than, say, carp.

The survival of the fertilised eggs is also lower and there is frequently a thick layer of dead eggs in the jar. This could be the source of various bacterial and fungal infections (to which these eggs are particularly sensitive) and therefore should be syphoned off occasionally.

Some protection from bacteria and fungi can be gained by using formalin in the incubating water at a concentration of 100–200 ppm. This concentration is not harmful to the eggs or larvae but will destroy the fungi and bacteria. The eggs will hatch in 24–36 hours in water at 20–24°C. During this period four or five formalin treatments may be given. The egg shell of these fish is very thin and breaks very easily. When egg infections occur the egg shells break much earlier than normal and the premature embryos collect at the bottom of the Zuger jar. The formalin treatment is indispensable to prevent this happening.

The fish larvae will be swimming actively 24 hours after hatching and swim to the surface vertically. Because of this the healthy larvae are not syphoned from the jar but allowed to swim out with the flow into a collecting vessel. The hatching of larvae can be shortened and synchronised with the use of enzyme treatment (Fig. 6.6). An industrial protein-decomposing enzyme is mixed into the incubation water at a concentration of 1:4000 and results in the complete hatching of the eggs in the jar in a few minutes. This treatment is performed when the first free swimming larvae appear in the jar.

Fry rearing

Feeding larvae are kept at a concentration of 100 000–500 000 in the large 200-litre jars. Larvae are ready for stocking into ponds on day 4 or 5.

The rearing of herbivorous fish larvae is similar to that of carp larvae. Preparation of ponds, and their stocking and maintenance, are based on the same principles as apply to the common carp. Practice has shown that these fish, although more difficult at some stages of the life cycle, can easily stand the intensive conditions of fry rearing. After the first month the results of fry survival are very favourable, equal to those achieved with common carp.

During the first month herbivorous larvae are always stocked in monoculture as all three species consume zoo-

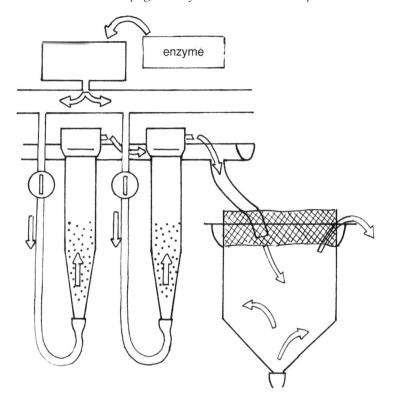

Fig. 6.6 By using proteinase enzyme, hatching can be accelerated and synchronised.

plankton (rotifers and lower crustaceans) during this period and would thus be competing with each other for food. After the first month the larvae will gradually become more specific regarding their nutritional requirements, according to their species. The silver carp filters algae, the grass carp eats duckweed and the soft parts of larger plants, whilst the bigheads remain filtering zooplankton and consuming colonial forms of algae.

If zooplankton is quickly used up (in 15–20 days) silver carp will more rapidly start to consume algae (as the adult does) but the growth rate is slower than if a zooplankton diet had been maintained longer.

It is important in the early stages of silver carp culture that the third plankton step is omitted. Large *Daphnia* cannot be consumed by the silver carp but the *Daphnia* will feed on the algae and hence be in competition with the fish for food. If there are large numbers of *Daphnia* in the pond prior to harvesting the fry, the pond should be treated the day before with insecticide (see Chapter 4.1 for details).

6.2   The propagation and rearing of tench (*Tinca tinca*)

Tench in mixed culture

Tench are a species of fish that does not respond well to intensive rearing in ponds. It cannot survive the intense competition for food that exists when they are grown with carp over long periods. Because of this the production of tench has declined in Europe although there are specialist markets (for example, in the ornamental and sport fish trade). Tench find living conditions most suitable in old ponds where the conditions are not favourable for carp (overgrown sites with rich muddy bottoms, characterised by releasing gases). In such conditions tench provide a good secondary crop to carp in extensive culture.

Traditional husbandry

In the traditional farming practices, the mixed wintering stocks are transferred to fattening ponds or marketable-sized fish production ponds in the beginning of the season where mature fish will spawn on the vegetation.

Significant numbers of the eggs and larvae will be eaten by carp and other species but a proportion of the offspring will reach fingerling size.

Semi-intensive husbandry

As a result of increased interest in juvenile tench, hatchery techniques have been improved for the supply of fry for intensive rearing.

The basic techniques described for carp are also applicable to tench but there are some specific differences for this species that should be highlighted.

The determination of the sexes is performed by observing the ventral fins of the fish. The edges of the female's fins are thin and straight, whereas the male's are much larger, thicker and curved (the male's fins are large enough to cover the vent completely). With well prepared broodstock the female has a large rounded belly (Fig. 6.7(a)).

Preparation for propagation can be performed after sex determination in small ponds or in bigger ponds with herbivorous species. Although wild spawning in deep ponds is not common, for safety sexes should be separated in April. Tench kept with carp females are not suitable for propagation as their egg development may have been suppressed. Tench require warmer water than carp to spawn and consequently spawning takes place later. The season usually lasts from mid-June to mid-July.

It is advisable to start the spawning programme with a group of 20–30 females and 20–30 males stocked into the hatchery. Hormone treatment is done once, using carp pituitary glands. For the mature tench, 9–10 mg (three whole glands) are used and injected into the belly (at the base of the ventral fin). Males and females are given the same amount of hormone. For large fish extra hormone is given (four to five glands (Fig. 6.7(b)).

It is advisable to place a medium in the spawning tanks that floats on the surface and provides shade. Fish will spawn and stick their eggs onto this material. In undisturbed conditions the females kept with the males should begin spawning 16–18 hours after treatment (at 24°C). After waiting for 10 minutes, stripping can begin. The maturation of the tench eggs develops periodically. Because of this the stock is stripped several times. Females should be stripped again one hour after the first to release the eggs that have matured during this time.

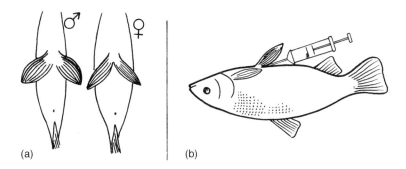

(a)   (b)

Fig. 6.7   (a) The pelvic fins of the male tench are large and rounded, whereas the female has smaller and thinner ones. (b) Injection site for mature fish.

Unlike carp and herbivorous fish where the eggs of one female are collected in a single bowl, the eggs of several female tench can be pooled together, since egg volume is quite low at first.

Males provide only a small amount of milt, despite having a pituitary injection. Because of this males are thoroughly dried and stripped on top of the bowl of eggs. If enough milt cannot be obtained by stripping the males, then one or two should be sacrificed and their testes removed and squeezed through a fine mesh. Two or three droplets of milt are sufficient to use with 100 mls eggs. Fertilisation is achieved by using the salt/urea fertilisation solution (1 litre of water containing 4 g saline plus 3 g urea).

Tench eggs are less sticky than carp so a tannin treatment is not really necessary. Eggs transferred to jars may stick somewhat on contact with fresh water, particularly if swelling is not complete. The rinsing period should therefore be reduced.

As the eggs stick in the jars, they should be stirred carefully and frequently to separate them. This will be necessary for the initial 10–22 hours' incubation. It is advisable to treat eggs with a malachite green fungicidal treatment, particularly with tench eggs when the fertilisation rate is less than 70 per cent.

Amongst farmed fish, the fry of tench and pike perch are the smallest, which must be considered when using net meshes to hold fry in the large Zuger jars. The non-feeding stage of the tench lasts for four or five days at 23°C, which is considerably longer than for carp or herbivorous fish. It is important to delay the stocking of feeding fry until all fry are swimming efficiently, otherwise survival rates will be very low.

The fry rearing is similar to carp. Because of their small size it is even more important to ensure that there is a large concentration of rotifers. Experience has shown that the slow-growing tench fry prefer higher animal protein in their supplementary feed (meat or fishmeal flours).

The fry rearing stage should be extended as long as possible to ensure that fry are strong at the time of harvesting. During harvest tench may bury themselves in the bottom mud. Many remain in hollows and footprints as the water level is lowered, therefore it is essential that water levels drop very slowly. The fry may swim out of the pond with the draining water at the very end of the harvesting operation, making this final phase extremely critical. If there are large numbers of the piscivorous insect *Notonecta* in the pond prior to harvest, it should be treated with insecticide the day before.

6.3   The propagation and rearing of pike (*Esox lucius*)

Breeding control of pike

In general, pike is the predatory fish of lakes and backwaters rich in wild cyprinid fish and aquatic vegetation.

Pike will breed in farm pond conditions only under exceptional circumstances.

The propagation of pike fry is considered only because they are produced in some hatcheries and small ponds where the hatchery building is empty at the beginning of the season. The fry are not generally reared on farms, but are stocked into natural waters.

Traditional breeding

The propagation of pike was achieved historically by catching fish on the spawning grounds and stripping the sexual products out on the spot (Fig. 6.8a). Eggs and sperm were mixed in a bowl and water added. Eggs were allowed to swell for 0.5–1 hour and then transported to the hatchery. The success rate for fertilisation would be 60–70 per cent.

The disadvantage of this method is that results depend on the natural spawning of pike and the season may last only for several days and the number of viable broodstock will vary annually.

Hatchery production

For a more secure breeding programme, pike should be propagated in the hatchery.

During the autumn, pike stocks selected for propagation are collected and held in wintering ponds (Fig. 6.8b). To keep condition, pike require a lot of food, consuming two to three times their body weight of trash fish until spawning time. If the food supply is insufficient, females become unsuitable for propagation in a short time. The ripening of eggs can be

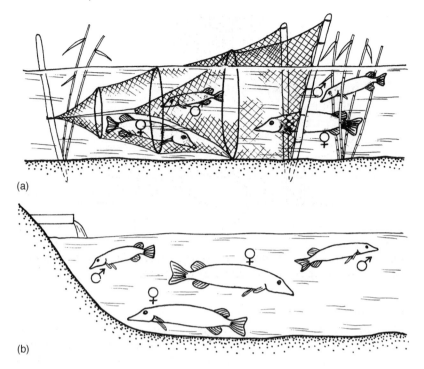

Fig. 6.8 (a) Broodstock pike are easily caught in fyke nets in the marginal spawning areas. (b) Broodstock selected in the autumn are held in wintering ponds where they are fed on trash fish.

promoted by using carp pituitary extract. Males are treated twice, and females once, using 3–4 mg of pituitary/kg body weight. In water temperatures of 12–16°C, stripping can be tried 36–48 hours later. If males fail to provide sufficient milt then the testes should be removed and pressed through a fine mesh. The milt obtained should then be poured onto the eggs.

Eggs of pike are very sensitive to shock so during stripping they should be carefully stripped down the side of the bowl and never allowed to fall onto the bottom (Fig. 6.9).

Fertilisation is achieved using a few millilitres of milt per litre of eggs. Fertilisation solution is then added. The fertilisation solution used for pike is stronger than that used for carp and is composed of 150 g urea plus 70 g sodium chloride in 10 litres of water. The fertilised eggs swell for 30 minutes. During swelling they are stirred very carefully as the eggs are very sensitive at this time. The temperature of the incubation water should be between 8–16°C, with an optimum of 14°C.

In the first two days the egg sensitivity is still very high, so

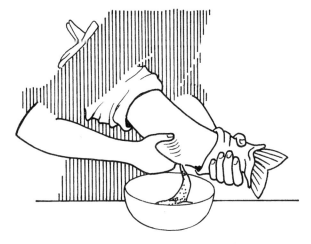

Fig. 6.9 Eggs are very sensitive to mechanical damage and should be carefully stripped down the side of a bowl.

the water flow should be slight to prevent damage. After this period, the eggs can be carefully stirred with a rod and the flow increased to 0.7–0.9 l/min which keeps the eggs turning slowly. Until hatching commences the water flow should be increased gradually as the oxygen demand increases until it reaches 1.5–2.5 l/min directly before hatching. The colour of eggs gradually changes during ripening, from yellow through greenish yellow to dark brown. This is due to the pigmentation of the developing embryo. A large percentage of unfertilised eggs is quite common, particularly with injected broodstock, thus the danger of fungal attack is very high. Malachite green treatments must therefore be given daily. The layer of dead eggs should be removed regularly.

The time of hatching is signified by the appearance of free embryos. One of the most sensitive larvae of farmed fish is that of the pike. After hatching, if the larvae are left to be turned over with the egg mass in the jar they will be easily injured. Hatching should therefore be done in very large vessels with large surface areas. Eggs are syphoned from the jar with a small amount of water and placed at the bottom of the large vessel in a thin layer. This should be done when all living eggs have hatched. (Note: sunlight and radiant heat accelerate the hatching process significantly.)

The freshly hatched larvae should be taken to holding cages. 8000–12 000 larvae are stocked into a net cage of 80–100 litres in volume. Strong, healthy larvae soon attach themselves to the sides of the net leaving the egg shells at the bottom. The egg shells can then be carefully removed with syphoning. If the larvae are disturbed during this process the fine protein strand that is used by the fish to attach itself to the net will be broken. If this happens, the fish then exhaust themselves trying to regain a hold and sink into deoxygenated detritus at the bottom of the cage. Large mortalities can result in this way.

After a long period of continued development these non-feeding larvae eventually detach themselves from the net and swim horizontally in the lower third of the net. Balance is achieved with rapid fin movement. The fish are now ready to feed. At this stage they are stocked into natural waters with a high proportion of vegetation, or they are reared in prepared ponds for a few weeks (Fig. 6.10).

Fig. 6.10   Feeding pike larvae are released in very small numbers over large areas of plankton-rich and weedy water to prevent cannibalism later.

Rearing in ponds and tanks

For ongrowing to the fingerling stage, small ponds or tanks can be used. The provision of food is the greatest problem. Pike larvae are much larger than cyprinid larvae, but full-size cyclops still represent a danger, so collected plankton should be filtered to ensure that the larvae are given only rotifers, small cladocerans and small cyclops. They will eat large quantities of plankton and grow very fast and this harvested plankton can only meet their food demands for a few days. Cannibalism appears very quickly in tanks and can be avoided only for short periods by providing an abundant food supply. From the second week plankton should be replaced by feeding finely chopped tubifex every three to four hours. Tank rearing should last for two to three weeks when fry should grow to 25–30 mm in length. For successful pond rearing very careful pond preparation is essential. In the cool early spring conditions plankton develops slowly so the ponds should be filled three weeks before stocking is planned. Fertilisation should be done during this period. Chemical treatment is useful only if large cyclops are present. After five days the chemical should be denatured and the pond should be inoculated with plankton in which cladocerans predominate. *Daphnia* can be included in this inoculation as they provide a good food source for the developing pike fry.

Fish survival in these ponds depends on the food supply and the availability of 'hiding places' (provided by aquatic plants). Pike prefer to hide behind plants and develop a hunting territory. If there is a lack of hiding places the fry are constantly moving and the risk of cannibalism is much higher. Cannibalism can also be avoided by harvesting the pond as soon as the plankton supply diminishes.

6.4   The propagation and rearing of zander (pike perch) (*Stizostedion lucioperca*)

Propagation

The zander or pike perch is a very aggressive predator. It does not respond well to intensive pond culture and hence should only be stocked in natural waters and extensive farms. Any zander that enter fry rearing ponds would cause extensive losses. Furthermore, zander require considerable care when handled and consequently many fish may be lost as a result of carelessness at harvest.

The propagation of zander differs in many respects to the techniques mentioned previously. It is not possible to strip eggs, and consequently it is easier and more successful to secure large numbers of fertilised eggs from natural spawning methods.

In waters where there is a significant zander population, sufficient eggs can usually be collected in the spring (Fig. 6.11). To do this a 'nest' is constructed out of willow branches and net secured to a line. Nests are constructed on a frame 200–300 mm wide and 300–400 mm long and placed in areas where the zander generally congregates (Fig. 6.12). Spawning starts in water at 10–12°C in early April. First,

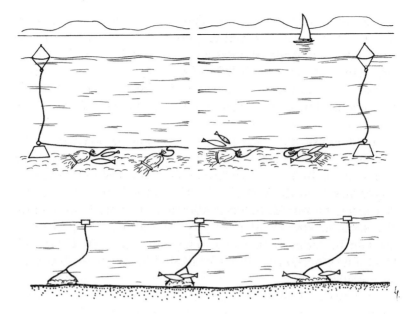

Fig. 6.11 Zander nests are attached to a large rope in natural spawning areas. Nests spawned on in wintering ponds are fixed individually on floats.

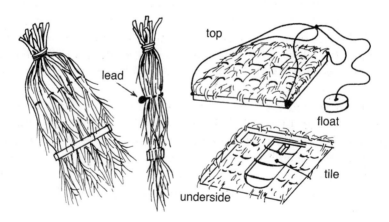

Fig. 6.12 Zander nests from natural waters are the shape of a brush, whereas in farms they are angular due to their wooden frame.

males clean the nests, and then females release their eggs. Nests covered with eggs are guarded by the males until the fry hatch. Nests should be checked every day and those covered in eggs should be transferred to the hatchery. As well as collecting eggs from natural waters, similar methods can be used in farm ponds. Pairs of similar sized fish are stocked into small wintering ponds. At the same time an appropriate number of nests are placed in the ponds. (Stocking rates are one pair/20–30 m².) Spawning commences within a few days. Zander are particularly fertile fish and a nest containing a few hundred thousand eggs may not be uncommon.

The fertilisation rate for zander eggs is excellent. The transport of their nests should be performed by covering them with canvas. In a multifunctional hatchery nests are placed in Zuger jars or in through-flow tanks in the hatchery, whereas in a specialist zander hatchery they are incubated under spray in a spray chamber. A fine spray is produced from an overhead sprinkler system. Zander eggs hatch in 6–10 days at 12–15°C. During ripening a daily malachite green treatment is necessary.

The eggs are treated in a malachite green solution of 1:60 000 for two to three minutes and then washed off. The eggs are quite resistant to the treatment and those contaminated with fungus turn blue in this strong solution. Eggs ripened in the 'spray' are regularly checked for maturation to avoid hatching in the chamber. A few eggs are taken into a small beaker. If they hatch within 30 minutes the nest is ready for hatching and should be transferred to the nursing ponds. Great care should be taken in equalizing the temperature of the nests, which should be tied to a suspended rope under water between two posts.

The non-feeding stage of the larvae is quite long (five to nine days) at 12–15°C. This period will be longer in colder ponds. During this period the fry continually move vertically in the water column and then slowly sink back to the bottom. Because of this characteristic zander larvae do not sink into the mud and can therefore be kept safely in natural ponds.

Other techniques have been tried in an attempt to keep larvae in the Zuger jars after hatching up to the feeding stage. In this protected environment there is less chance of damage to the stocked fry which may be caused by variable spring weather (e.g. fish can be held for several days if there are strong winds over the ponds).

The fry can be fed with hard-boiled egg mix which starts them feeding in the hatchery, thus giving them an advantage when they are stocked out.

Sensitivity of the zander fry

It is well known that zander fry are very sensitive to temperature shock particularly when the larvae are very young. Because of the small size of the fish on stocking out the results of temperature shock may not immediately be apparent.

During stocking the temperature difference between the hatchery water supply and the receiving pond should be carefully and slowly balanced. If this process is rushed the effects may be fatal. Acclimatisation should be performed over a period of 30–60 minutes.

Pond preparation

Because of the small size of the zander the appropriate pond preparation and multiplication of small-sized food organisms is essential. The zander fry have an instinct to hunt at a very early age and are able to catch the nauplius stage of cyclops which move very quickly. Other fish species are unable to catch these organisms.

This relatively high speed is insufficient to protect the zander from the adult cyclops, so during pond preparation the chemical treatment eliminates cyclops and the rotifers can be stimulated by the use of fertilisers. As plankton development is much slower in the early spring, ponds should be filled three weeks before the planned stocking.

During the long preparation period it is possible that the cyclops may multiply again. Because of this a few days before stocking (or in the case of pond hatching, a few days before hatching) the plankton population should be assessed. If cyclops have reappeared then the chemical treatment should be repeated (as the treatment does not affect either the eggs or the larvae).

As soon as the fry start feeding, the ponds should be inoculated with zooplankton. In spring the bigger *Daphnia* is stocked instead of *Moina*. At the same time the pond is flooded further and inoculation continued. With this intensive management of the food organisms the survival rate of the zander will be greatly increased, as the effects of cannibalism are greatly reduced.

When the plankton population is highly concentrated the small zander grow very quickly. After one month the fish are

40–50 mm long. At this stage the appetite of the stock is so great that the pond cannot supply enough food and cannibalism quickly becomes a problem.

To prevent this, harvesting should be commenced. The harvesting of such small sensitive fish, however, is not easy. Even with the careful use of very fine nets mortalities will result. Because of this they should be caught in a trap during the drainage of the pond and this is best performed during daylight hours. Considerable care should be taken to ensure that temperatures are carefully balanced when transporting these fish.

After the appropriate preparations, zander fry can be transported large distances using tanks with fine oxygen diffusers.

6.5   The propagation and rearing of wels catfish (*Silurus glanis*)

The wels catfish stands out as a most suitable fish for intensive polyculture farming conditions. Its temperature and oxygen demands are similar to that of carp when feeding conditions are favourable, it is reasonably resistant to disease and can withstand the shock of handling during harvesting. It has a big appetite and grows fast. Wels catfish consume organisms in the pond that other fish do not utilise (large aquatic insects, tadpoles, wild fish, sick fish, etc.). Unlike the zander it will search for fish only if it cannot satisfy its appetite in other ways.

Propagation

Pond husbandry techniques for wels catfish were developed in Hungary. A propagation method was developed many years ago that is still used in its original form on a number of large farms. Only recently has intensive technology been developed for this species.

In both methods, breeding stock are prepared the same way. For propagation fish with a body weight of 6–12 kg are the most useful.

These fish are kept in wintering ponds with sufficient supplementary feed. In one 600–1000 m^2 wintering pond

60–100 fish are prepared. Trash fish is stocked into this pond at a density of three to four times that of the total weight of the broodstock.

If possible, trash fish such as roach or rudd should be used rather than common carp or goldfish as they are less susceptible to bacterial infections. This presence of trash fish is particularly important in the spring. Egg development requires large amounts of high quality nutrients at this time. According to estimates, 30 per cent of the total trash fish are consumed in two months of spring. In the first half of April the broodstock are separated to prevent wild spawning. Certain diseases may develop in these wintering ponds, particularly white spot. These should be treatable as long as symptoms are recognised early and the disease identified by a veterinary surgeon.

Natural spawning

Natural spawning can be promoted in small fish ponds. The fish farmer reproduces the environmental conditions in the pond to represent those that the fish might find in the wild. Normally wels will make a nest in the roots of willow trees, and therefore to encourage the female to lay her eggs the farmer builds a nest of fixed rods in a tent shape into which is woven the washed roots of the willow (Fig. 6.13).

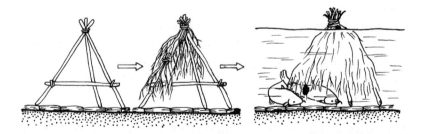

Fig. 6.13 Willow roots are attached to the wooden frame to construct the catfish nest. The pond is flooded and finally the broodstock are introduced.

Wels spawn in pairs. Good results can be obtained only if the male and female are similar in size. If this is not the case, fighting may occur and the weaker fish may die within a few days. Any spawning that does take place will be of poor quality.

Sex determination was a particularly difficult aspect of wels cultivation for a long time. The sex of the fish is determined by a plump belly, the shape of the genital papilla (Fig. 6.14) and the shape of the head. Fish that cannot be readily sexed should not be used for propagation.

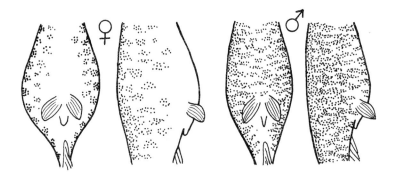

Fig. 6.14 Well prepared broodstock are needed for successful hatchery propagation. Sexes are distinguished by their genital papillae and swollen abdomen.

Spawning commences when water temperatures reach 20–22°C for several days. Small ponds or wintering ponds chosen for spawning are filled a few days prior to the planned spawning time, which is usually during the first few weeks of June.

Before filling, the nests are fixed to the bottom of the wintering pond by stakes. If the water temperature is suitable, spawning starts a few days after stocking. It happens during the night and in the early hours around dawn and is recognised by turbulence and splashing. Following the spawning the males will guard the nests for several days. Nests should not be moved after the second day following spawning as the embryo may be broken out of the weakened shell. Nests should therefore be moved early to concrete ponds in a dismantled farm or into net cages. Nests can also be taken complete to ponds where the rearing of fry is planned. Eggs will hatch in four to seven days, depending on temperature.

Newly hatched wels larvae are very weak and attach themselves to the willow roots by a thin protein 'string'. For two or three days after hatching the nest should not be disturbed as the fry may be dislodged and fall into the bottom mud where they would perish. At this stage cyclops can cause large losses to the fry.

Surviving fry have considerable appetites and they grow fast. They will catch rotifers and middle-sized species of zooplankton with their large mouths. Ultimately, cladocerans, cyclops and mosquito larvae will be eaten. This large appetite can cause problems when the fish are reared into wintering ponds as the natural food supply soon diminishes and the fry begin to starve.

When the cyclops disappears from the pond the larvae of unicellular parasites can multiply quickly and attack the wels fry. If white spot appears only a very weak malachite green treatment can be administered (0.1 mg/l) and fish should be harvested a few days later. The infected fish are too weak to withstand a normal or repeated malachite green treatment and will perish.

Fry are ongrown by stocking them thinly with different age groups of carp. It is essential, however, that the carp stocking density is low, and that there are plenty of hiding places in the plant life. If the opposite occurs, the young wels will be consumed by the carp.

Propagation techniques in the hatchery

The equipment used for spawning and rearing wels in the hatchery is the same as for other fish. However, as there is much less room in the tanks for the broodstock, they can become very aggressive and they will damage each other. The main advantage of using hatchery facilities is that large numbers of fry can be produced in a relatively small space. To keep fish from fighting in the holding tanks their mouths are fastened closed. The nasal bone is drilled and the mouth tied with twine (Fig. 6.15). Before tying, the fish are anaesthetised and injected with hormone at 4–5 mg/kg body weight (diluted carp pituitary extract) to provide an injection of 1–1.5 mls. Ten per cent is used as a preliminary injection with a 12-hour time between. The injections are given into the muscle and backflow is prevented by efficient massage of the injection site. Broodstock are kept at 23–24°C and after 12 hours eggs are released and can be stripped. Historically it was thought that wels could not be stripped because of their anatomy (with a muscle forcing the oviduct closed). However, when the fish

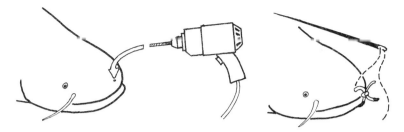

Fig. 6.15 To prevent fighting the mouths of the broodstock are sewn up with twine.

are anaesthetised the muscle relaxes and the eggs can be freely stripped.

The stripping and fertilising process differs slightly from other species. Milt is usually collected by pipette; however, if there are large numbers of eggs the male is opened up, the testes removed and passed through a fine mesh. If this operation is performed carefully, the cut can be stitched up. Most fish (80–90 per cent) will survive such an operation although usually these males are sacrificed and sent to market.

Sperm from two fish should be provided for each batch of eggs. The eggs and milt, however, are not mixed in the 'dry' form. Mixing only occurs at the moment when the fertilisation solution is added by quickly spinning the bowl and its contents. 20–30 mls of fertilisation solution are added to 100 g of eggs. The fertilisation solution contains 3 g salt per litre of water.

Eggs can only be stirred after the fertilisation solution is added. Wels eggs are very sticky. Swelling is only allowed to develop for a few minutes before the complete batch of eggs (100–200 g/glass) is poured into the Zuger jar. The eggs will stick to each other and then as a result of contact with fresh water will swell to many times their original volume.

During the early stages of development the wels eggs require very little oxygen and can stay stuck together for 10–12 hours. During this time the water flow percolates between the eggs to provide sufficient oxygen. As development continues so does the oxygen demand and the danger of fungal infection increases. The eggs must now be 'unstuck'. Ten to twelve hours following their placement in the jars the eggs are treated with a protease enzyme at a concentration of 0.04–0.05 per cent for four to five minutes. During treatment, eggs are stirred with a plastic rod so that the enzyme solution can reach all parts of the egg surface. The enzyme digests the protein layer that causes stickiness, after which the eggs separate freely in the flowing water (Fig. 6.16). They hatch in three days at 22–24°C. Daily malachite green treatments are very important. In the stages prior to hatching the eggs increase in size as the moving embryo enlarges the weakening egg shell.

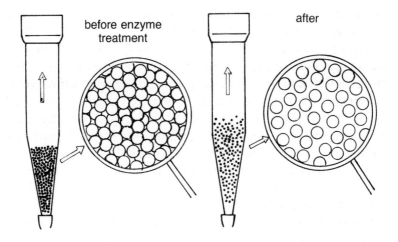

Fig. 6.16 The stickiness of wels eggs is eliminated using an enzyme treatment. Following this they will separate.

An interesting characteristic of the swollen egg is that the infertile eggs turn white and ultimately lose their egg shells and sink to the bottom of the hatching vessel. When it is time for the healthy eggs to hatch (i.e. when the first few larvae appear in the jar) these good eggs are syphoned out through a wide plastic tube into bowls, leaving the dead eggs behind in the jar. The eggs removed for hatching should be kept submerged under a thin layer of water.

The hatching of the egg mass takes place five to ten minutes later in the bowls (on tapping the side of the bowl the larvae start to move).

Wels larvae are very fragile and can be easily injured. The larvae should be kept in the nets used for pike culture. For the first few days after hatching the yellow larvae hardly move, remaining on the bottom. During this time it is important that they are prevented from gathering in cavities in the holding net where they would damage each other. On the second or third day, the larvae gradually turn grey and move out into the darker corners of the holding net. During the next one or two days they are able to feed, and this stage becomes identifiable by the larvae swimming along the edge of the net at the surface.

Larval rearing

The feeding larvae can be reared further in several ways. They can be reared in plastic tanks with a continuous flow through for a few critical weeks. The tanks need continuous management for cleaning, feeding, disease control, etc. The gluttonous wels larvae can only be satisfactorily fed on tubifex larvae. Waste and faecal matter should be removed several times throughout the day. As long as feeding rates are sufficiently high cannibalism should not be a problem. After a few weeks the wels larvae will have reached 20–30 mm in length and will be too large for a tank environment. They are now suitable for stocking into the less favourable pond environment.

If feeding larvae are reared in ponds there will be dangers from preying cyclops as well as from a reduction of the zooplankton population. Better results can be achieved if surplus fry of tench and silver carp are stocked at the same time. They will be grazed by the wels and good growth rates will be achieved. Common carp fry, however, would compete with the wels which may result in some carp surviving. The small pond rearing stage should only last for a few weeks (three to six) to ensure sufficient feed availability. If this period is extended white spot disease may occur. Because of this, after rearing the fish for one month the small wels are stocked as secondary species in large ponds. They are stocked at 500–1000 fish/hectare. At this stocking rate they should not endanger stocks of other species and, in fact, will significantly increase the total yield of fish without increasing the supplementary feed rate.

The small pond rearing phase can be omitted but with greater risks to the stock. Instead, large ponds of several hectares can be prepared using chemical treatment. Since large losses can be expected, 5000–25 000 wels larvae are stocked per hectare and fed appropriately. The stock of other species is reduced significantly ($50–100/m^2$). Cyprinids can grow well in these favourable conditions and quickly become too large to be caught by the wels, so a further stock of common carp fry is introduced in the middle of the season to provide the wels with food into the autumn.

Recent developments in the culture of this species involve the rearing of fry in tanks using artificial food specially formulated for catfish.

6.6   The propagation and rearing of goldfish (*Carassius auratus*)

The wide variety of different types and forms of goldfish all have the same ancestor – the Chinese crucian carp (*Carassius auratus*). The present varieties have been produced by very resolute selection work over a thousand years. The crucian carp is a very hardy fish found in still water. It grows much slower and to a smaller size than its relative, the common carp and it is widespread across the cold/moderate climates as well as the tropical areas of Asia. An interesting feature of the fish is that its populations living in Far East Asia reproduce in the usual sexual way with an approximate male:female sex ratio of one to one, while in Middle Asian regions the closely related silver crucian carp populations that arrived in Europe by a gradual migration exhibit a unique mode of reproduction. These populations consist only of females and reproduce 'gynogenetically'. The females mate with males of other cyprinid species and the sperm of these males initiates the development of the eggs but does not contribute to the genetic material of the developing embryo. Thus the offspring are not hybrids but inherit only the genetic material of the 'mother'.

The ancestor of the goldfish is the normal Far East Asian variety. The red (gold) variety evolved by selection over a thousand years but in more recent times a large number of different variants have been developed by fixing several mutations. The longest tradition of breeding is found in Japan and China. The goldfish's popularity has not decreased; it has also become established as a domestic pet, with some particular specimens living for several decades with the family.

The goldfish's popularity results from its beauty, richness in shape and simplicity for keeping. It reaches maturation in a fairly short time, with males being capable of reproduction after one year and females becoming mature after a year in warm climates or after two years in colder areas. Goldfish are very easy to keep as they thrive either in pond or aquarium.

They tolerate a wide range of temperatures, especially where this alteration occurs gradually. They can live at temperatures from almost 0°C to 30°C although the optimum temperature is 20°C. Fish kept in garden ponds may start spawning at temperatures around 15°C.

Goldfish are not fastidious about food. They eat all lower invertebrates and small animals that are found in natural waters, seeds of plants and fragments of organic material. In artificial environments their most common nourishments are boiled rice, peas, tubifex, mosquito larvae, daphnia and high nutritional quality granulated or pelleted feeds. Among special feeds for goldfish are those containing carotene, which are popular because they reinforce the colouring of the fish.

The colouring of the fish depends on several factors. One of the most important is genetic propensity. There are strains whose populations transform very quickly from the wild type brown colour to a preliminary pale yellow then finally to flame red. Others change more slowly over a longer period whilst a certain proportion retains its brown colour. These brown ones may produce red offspring but the ratio of browns will increase over the subsequent generations. Without stringent breeding and selection the stock deteriorates and after a few generations the brown wild type will predominate. In addition to genetic inheritance other factors such as sunlight and diet also affect coloration. In sunny ponds the colour of fish is more intense and the coloration develops faster. In aquaria where goldfish are usually kept in darker places colouring might not develop at all and the fish retain a pale yellow colour.

A final important factor affecting colour is the accessibility of plant food. Goldfish, especially in dense stocks where little protein is obtained by the individual, show a preference for feeding on sestonic or planktonic algae, preferably green algae. In this case growth is slower but the colour is more vivid.

Natural reproduction

In garden ponds in Europe reproduction occurs during the first warm days of spring. It is signalled by a change of behaviour of the fish. They excitedly swim up and down the pond with males following the females, seemingly chasing them. Females with rounded bellies full of thousands of eggs attempt to escape from molestation by the males, at the same time searching for an area in the pond with dense vegetation. The spawning and laying of eggs usually commences at dawn.

Each mature female ready to release eggs is followed by a group of males which then swims into dense vegetation where intensive splashing takes place and the release of large numbers of eggs and milt occurs. The released eggs stick to the leaves of plants within a few seconds irrespective of whether fertilisation has been successful. Provided fertilisation has occurred the embryo develops within the egg. The alevins hatch in a few days and start their individual lives. The alevin of goldfish is similar in size to a newly hatched mosquito larva. This therefore makes it suitable food for the adults. Spawning goldfish parents often prefer to eat the new-laid eggs and hatched alevins! It is no wonder that goldfish reproducing in small size garden ponds (or in aquaria) do not produce many (if any) offspring. Human intervention is required to increase the chance of survival.

Before providing further details about spawning, some other aspects of goldfish keeping should be reviewed. After reproduction in the spring, the summer provides warmer weather when the most important activity of the fish is feeding. They consume everything that is suitable including waterborne invertebrates, airborne insects falling into the water, plant seeds and subsidiary feeds given by their keeper. As a result of this abundant food they show accelerated growth and prepare for the next reproduction. In favourable conditions they may spawn several times in a season as a result of the change in water conditions, particularly following heavy rains. Later during the autumn they accumulate appropriate amounts of energy to see them through the winter.

Without providing detailed descriptions of the large number of recognisable strains of goldfish (which can be found in specialist books) basically two groups of goldfish varieties can be distinguished. The first group consists of the simpler, wild type – single-tailed, long shaped, fast moving varieties (red common, comet, shubunkin, wakin, etc.) while the second group consists of the clumsy, veiled tail, delicate eyed fish (telescope-eyed, lionhead, bubble eye, etc.). The varieties of the first group can easily be kept in garden ponds or even in larger ponds. Since they are capable of escaping from predatory fish, birds and from frogs (younger age group) and are able to search for enough food they tend to grow quickly. They can also be kept in polyculture together with other cyprinids. The varieties of the second group require more controlled conditions such as aquaria, or small tanks where more demanding nursing (regular feeding, regular change of water) can be applied. Additionally, there are differences between the two groups in terms of overwintering requirements. While in the first case the fish can easily be wintered outside under ice, the varieties of the second group demand warmer temperatures. These must be wintered in cool and dark tanks or at room temperature in aquaria. In this latter case the fish has to be fed during winter.

Propagation of goldfish

As previously mentioned, goldfish parents may eat their offspring, therefore the successful rearing of larvae requires strictly controlled management. A fairly simple method of propagation is to leave the fish to spawn in the pond. In this case a spawning surface should be provided (aqueous plants with fine fibres or a non-rotting plastic filament – something onto which eggs can stick and develop further).

To avoid cannibalism by parents the larvae should be removed after spawning or, more simply, the medium with attached eggs can be removed and placed into another tank prepared for the incubation of eggs.

How can the spawning of fish be promoted? It is not easy in small ponds. The best method involves stocking males that show signs of being prepared to spawn (tiny white bumps on the head and around the pectoral fin) and females full of eggs (rounded abdomen) into a pond with no vegetation and with a bank of shallow water of 10–20 cm depth. When the water temperature reaches 13–15°C and periods of sunshine are more regular, larger amounts of aqueous plants should be placed into the shallow area in the early morning hours, distributing the material evenly and loosely. Fish feel the presence of a spawning surface and spawning starts, possibly lasting for several days. If spawning is successful lots of eggs can be observed by the naked eye, stuck to the surface of the plants. The plants which are covered with the eggs are then removed.

If spawning needs to be controlled over a short period of time (when there are plenty of broodstock to propagate to gain a high amount of eggs) goldfish can be stimulated by using one injection of pituitary hormone to spawn synchronously. One pituitary of carp (of 3–4 mg dry weight) can be used for four to five mature, average-size females or males. Treatment should be done in the early evening hours, stimulating spawning in the early hours of the following day. Pond spawning onto aqueous plants can be successfully achieved, especially in the case of the less delicate varieties of fish. The more sensitive and delicate varieties can be propagated in big plain tanks where the probability of damage is lower, the spawning is easily controllable and the fish can be nursed and fed.

In such environments not only spawning medium but the water itself can influence the time of spawning. While in small ponds water cannot be changed (this is normally dependent on the weather: warming up by sunshine and dilution by rains), in tanks we can improve the chances of spawning by changing the water, adding a few buckets of rain water and/ or hot water. Goldfish easily detect a 2–3°C change in the temperature and this is usually enough for the stimulation of spawning (Fig. 6.17). It is important to remember that these stimulating interventions should be done in the morning hours to gain a spontaneous ovulation during the following morning.

Pituitary hormones can also be successfully used for the

Fig. 6.17   Goldfish spawning in small tanks.

stimulation of ovulation in these 'fancy' varieties of goldfish. The dose is similar to that mentioned earlier (approximately one quarter of a carp pituitary is used for the injection of a female). The hormone solution is prepared by powdering a dry carp pituitary in a porcelain mortar then diluting in a 0.65 per cent sodium chloride solution by adding 2 ml to one gland. Each goldfish is then injected using 0.5 ml of the prepared solution. This solution is carefully injected into the abdominal cavity by using a (preferably thin) hypodermic needle while keeping the fish upside down in the hand. Pituitary treatment is not an intervention without dangers. Fish that do not respond to the treatment may die in the next three or four days from the degradation of non-ovulated eggs retained in the ovary, which poison the fish. To avoid this, a specimen of rare variety or higher value should not be treated like this; instead, induction of natural spawning should be tried. Between two spawnings, fish should be fed intensively by adding natural food in high proportion (collected zooplankton, mosquito larvae, etc.).

A successful spawning may result in several thousands of eggs. Some of the well-prepared fish may produce as many as 10–15 000 eggs. The fertility of the eggs depends on the activity of the males, and the quality and quantity of sperm cells. In the case of fish varieties where skills of fast swimming and good sensitivity of the eyes have not been lost during centuries of breeding (i.e. the hardy varieties) an 80–90 per cent fertilisation rate can be gained, even in natural spawning conditions. However, in the case of the 'fancy' types, the disproportionately big eyes or the veiled tail may prevent active motility and, thus, effective fertilisation. Natural spawning achieves very low fertilisation rates (30–40 per cent) of these types.

A very useful improvement may be to strip the ovulated eggs from the females and fertilise artificially. Eggs can only be stripped when they are in the fluid ovulated state. A fish, when spawning, produces flowing eggs, and consequently it can be stripped. Only fish apparently spawning should be tried. Eggs are stripped into a small dish after drying the abdomen of the female, thus avoiding water getting onto the eggs, since water can activate the eggs prior to fertilisation and make them unfertilisable.

Males are then immediately stripped by using a fine dry pipette. Sperm also dies prior to fertilisation if equipment is wet. Two or three droplets of milt are added to the dry eggs and mixed well. Then 1–2 ml of pond water (or dechlorinated tap water) are mixed with the eggs. Care should be taken to ensure similar temperatures in both the water where the goldfish were kept and that used for fertilisation. The effect of the water is that fertilisation is stimulated and occurs in a few seconds, resulting in a 95 per cent success rate. Water not only activates but also causes the egg surface to become sticky. If kept in water eggs will conglomerate in a few seconds, stick together and ultimately perish. To avoid this, fertilised eggs should be spread over a surface, preventing them from touching each other. Aqueous plants, fine plastic fibres, or pine tree branches are suitable for this function.

If many eggs have been produced and there is suitable experience and equipment, goldfish eggs can be incubated in Zuger jars. The eggs are treated using the procedure applied to carp (salt-urea solution is used for fertilisation; 40 g sodium chloride plus 30 g urea are dissolved in 10 litres water). Eggs of goldfish hatch in three or four days. Constant temperature should be ensured with an optimal oxygen level of 5–6 mg/l. Alevins do not feed for the first four days. Their size is similar to that of carp fry (a little smaller) but their feed is the same. If there are large numbers of larvae it is reasonable to rear them in small ponds. In this case the pond is prepared similarly to that in carp fry rearing. The 'fancy' varieties are reared in tanks for a couple of weeks using artemia (brine shrimp). This avoids having to stock the fish into the pond at a very small size, as they would easily become prey to frogs and aqueous insects.

If many thousands are to be reared, ponds may be used to stock the fish larvae but they should be protected using a frog-proof fence!

6.7   The propagation and culture of African catfish (*Clarias gariepinus*)

As the name suggests, this species has its origins in Africa where it is well adapted to the extreme African climate with its long-lasting hot and dry periods. This species has a unique accessory breathing system, a special organ in the gill cavity with a large surface rich in blood vessels that enables the fish to absorb oxygen from the air. When the level of dissolved oxygen in the water is low, the catfish swims to the surface and fills its gill cavities with air. After sinking back to the bottom, the fish is able to store the air for long periods. The fish then absorbs oxygen from the air stored in this organ. *Clarias* is a strong, undemanding species that has adapted well to the unfavourable conditions often found in the 'temporary' waters of Africa. It can live in marshy, salty conditions, rich in organic matter and toxic gases. Its tolerance to ammonia is several times higher than that of most other species. *Clarias* also tolerates high stocking densities and will eat a wide range of foodstuffs. All of these qualities makes this species well suited to aquaculture, particularly in intensive recirculation systems. The main drawback for cultivating this fish in temperate climates is that it requires warm temperatures. Below 15°C it becomes very prone to disease and will die.

In addition to the many qualities listed above, *Clarias* is easy to propagate, reacts well to hormone treatment, and with its ability to rapidly develop eggs it can be spawned several times a year.

This species is well known in Africa, but for aquaculturists it was only recognised in the 1950s when Dutch researchers recognised its excellent qualities. Originally the fish was used in warm water effluents such as power-station cooling water and geothermal waste waters. In the summer the fish can be reared in outdoor ponds. With the development of sophisticated recirculation systems, *Clarias* is now successfully reared in large numbers. Production costs are low and the boneless meat is favoured by many.

Propagation

Clarias reaches sexual maturity quickly and it is possible to use fish weighing between 400 and 600 g (8 months–1 year old) ongrown in intensive systems. The fish reacts well to both carp pituitary and GnRH (see page 41). The usual dose of pituitary is 3–3.5 mg/kg body weight for females and 2–2.5 mg/kg for males (Fig. 6.18). When using GnRH analogue, the dose should be 15 mg/kg body weight plus dopamine antagonists (10–15 mg of metoclopramide or 5 mg domperidone). No pretreatment injection is needed to induce ovulation, which occurs 10–12 hours following hormone treatment at 26°C. Egg collection is shown in Fig. 6.19. Because of the anatomy of the

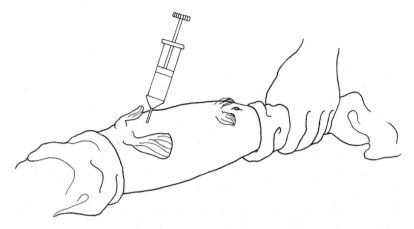

Fig. 6.18   Hypophysation of a ripe African catfish (*Clarias* sp.).

Fig. 6.19   Egg collection following hand stripping of *Clarias*.

testes it is not possible to strip sperm from the males. The males thus have to be sacrificed and the testes surgically removed. Semen can be squeezed out of the testes on to the egg mass and water used for fertilisation.

A unique characteristic of the eggs is that a disc-shaped sticky area develops around the micropyle that 'glues' freshly stripped eggs together in the hatching jars. This stickiness is, however, only temporary and disappears during incubation.

A few hours following fertilisation, eggs are dispersed by gentle stirring and they develop quickly. Hatching takes place in 24 hours at 26°C. The catfish larvae are initially sensitive to low dissolved oxygen levels and infections by bacteria and fungi, so care has to be taken at this stage.

Within 3–4 days larvae can be fed on *Artemia* nauplii, or decapsulated *Artemia* cysts. Within 3–4 weeks the breathing organ has developed and they can be cultured with increasing intensity. Recirculation systems can rear $100\,\mathrm{kg+/m^3}$ using complete diets (35% protein). Research is currently looking to use cheap protein sources to allow the fish to be produced more profitably.

6.8   The culture of sterlet (*Acipenser ruthenus*)

The only Central European sturgeon species that spends its entire life in fresh water is the sterlet (*Acipenser ruthenus*: see Fig. 6.20). It belongs to the group of ancient Chondrostean fishes. In the wild this species requires clean river water, where it feeds primarily on invertebrates and organic detritus that is found in the mud. It is extremely sensitive to polluted water. Sterlet produces excellent meat quality, and also has value as an aquarium fish.

The spawning process begins in early May, when at a water temperature of 10–12°C sexually mature fishes gather on the spawning grounds: the sand beds of fast flowing rivers.

Artificial propagation

Unfortunately, when the spring is warm, the spawning period of sterlet is very short (restricted to a week). The spawning season can be extended by farmers if they keep the females cool and tranquil.

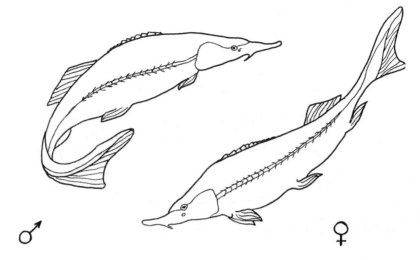

Fig. 6.20   Broodstock of sterlet.

According to the literature, dry pituitary glands from other sturgeon species (the large migrating species such as Beluga, Sevryuga, Russian sturgeon, etc.) are most suitable for the hormonal stimulation of mature sterlet brooders. The best results have been achieved using these. However, the major sturgeon culturing countries (i.e. the former Soviet Union states and Iran) closely guard their monopoly in caviar production, thus the export of sturgeon pituitaries is not generally possible.

If sturgeon pituitary gland is available, the usual dosage for inducing ovulation is 2.5–4 mg per kg of body weight. Males and females both have to be injected (Fig. 6.21a). Better results are achieved if the females receive a pretreatment injection (10%) 24 hours before the decisive dosage.

Stripping can be started 2–3 days after the treatment depending on the temperature, when a few black eggs appear attached to the bottom of the holding tank. The fish should then be removed from the water, held in one hand by the back of the head and the base of the pelvic fins; in this position ripe black eggs should emerge from the genital aperture in a pearl-like chain. If only a few eggs appear, ovulation has not yet been completed, or only partial ovulation has been produced.

Ovulated eggs are released freely into the abdominal cavity, as sturgeon ovaries do not have an external 'lamina' like those found in cyprinids. Because of this, eggs cannot be stripped.

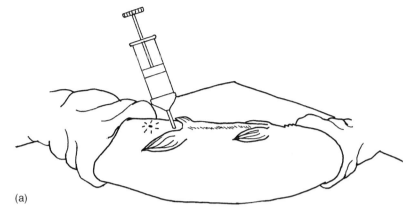

(a)

Fig. 6.21a Injection of sterlet broodstock.

They have to be removed following a surgical opening of the females (Figs 6.21b and c).

Sterlet milt is a slightly cloudy, opalescent fluid, which can easily be wasted during the handling of males because the stressed, splashing fish can flex their abdominal muscles and squeeze it out. If the volume of sperm is insufficient, one of the

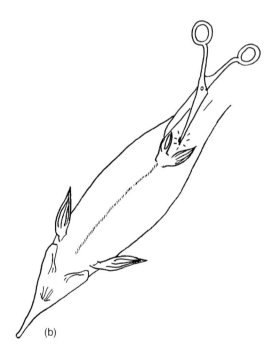

(b)

Fig. 6.21b Surgical opening of the body cavity.

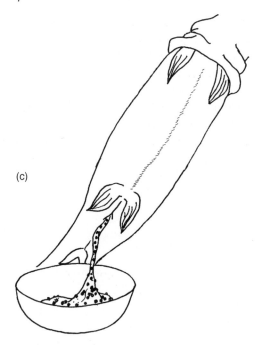

(c)

Fig. 6.21c   Collection of eggs following surgery.

males can be sacrificed, its testis removed, and squeezed through a mesh fabric.

During fertilisation, special attention should be paid to the dilution of milt. Since sturgeon eggs possess numerous mycropyles, 'polyspermy' can occur when undiluted milt is used for fertilisation, meaning that several spermatozoa can penetrate the egg. The consequence of this phenomenon is irregular development and death of the embryo. If the sterlet milt is diluted approximately 50 times, this can be avoided.

According to the literature, the pituitary gland of cyprinids can also be used for the propagation of sterlet, but the dosage has to be increased several times to produce a dosage of 8–15 mg per kg of body weight.

Because of the difficulties of 'sturgeon' pituitary availability and the requirement of such high dosages of 'carp' pituitary, gonadotropin hormone releasing hormones (GnRH analogues) can also be successfully used. In the case of sterlet, Gn RH is effective alone without the need of the dopamine antagonist for the hormonal stimulation of mature females. The effective dosage is unusually high: 50–100 µg per kg of body weight has to be injected into the abdominal cavity to stimulate ovulation. At an incubating water temperature of 13–15°C and 100% oxygen saturation, levels of about 80–100% of the broodstock should respond well to the treatment. The percentage of partially ovulated fish is generally low.

To 'de-sticky' the eggs, a starch solution (200–250 g of household starch diluted in 1 litre of 12–15°C water) is used as they are being water hardened (Fig. 6.22). The stickiness disappears gradually during the $1\frac{1}{2}$-hour process. When the fertilisation rate is high the eggs are less adhesive and are less likely to stick together.

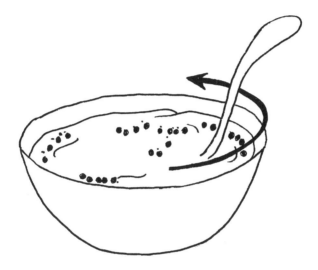

Fig. 6.22   Treatment of sterlet eggs using starch solution.

In these cold temperatures sterlet eggs develop slowly, requiring 5–7 days for incubation. During this time the eggs have to be treated daily with malachite green to avoid fungal and bacterial infection.

Larval rearing

During hatching the large and agile larva frees itself from the egg-shell, swims out of the hatching jar with the water flow, and can easily be collected in a larval trap attached to the outlet. In the non-feeding period, giant incubators (as described at the end of Chapter 3) are suitable for larval rearing. A strong flow through the rearing tank and aeration of the water are both essential during the early part of their development as the larvae need to gather in the flow during this period.

The onset of feeding is indicated by the absence of the black spiral excrement (known as 'intestinal tar') in the posterior part of their gut. Initially fish are fed on small worms (chopped tubifex), filtered zooplankton, or newly hatched *Artemia* nauplii. As they develop they can be fed with a high-protein diet such as trout feeds. The fish grow fast on this high quality food and reach 5–7 cm size at the age of 6 weeks. From this time the sterlet fry can be reared in intensive systems on formulated feeds.

Appendices

1. Reproduction Biology of Pond Fish
2. Hatchery Propagation of Pond Fish
3. Fry Rearing in Farm Ponds
4. The Yield of Carp (in Monoculture) in Farm Ponds
5. The Yield of Carp (in Monoculture) in Farm Ponds: Semi-intensive Management
6. Yield of Fish Ponds under Extensive Management (Polyculture)
7. Yield of Fish Ponds under Semi-intensive Management (Polyculture)
8. Long-term Treatments in Pond conditions (without Water Exchange)
9. Rapid Treatments of Fish in Tanks

Appendix 1. Reproduction Biology of Pond Fish

Specification		Unit	Common carp	Chinese carp	Tench	Pike	Zander	Wels
Age	Male	Years	2-3	4-6	2-3	2-3	2-3	3-4
	Female	Years	4-5	5-7	3-7	3-4	3-4	4-5
Weight	Male	kg	3-4	3-7	0.4-2.5	0.5-2.0	0.5-2.0	3-7
	Female	kg	4-5	3-10	0.75-3.0	1-5	0.75-3.0	4-12
Sexual maturation — Length	Male	cm	25-30	40-90	25-30	20-30	25-30	50-70
	Female	cm	30-40	40-100	25-30	30-40	30-40	50-70
Spawning	Month		IV-VI	V-VII	V-VI	II-IV	III-IV	V-VI
Water temperature		°C	16-22	21-25	22-24	6-12	10-12	22-24
Number of eggs gained from 1 kg of weight of females		Thousands	100-200	40-80	80-120	20-45	150-200	10-48
Number of eggs gained from a female		Thousands	200-1500	200-1500	40-100	17-220	100-300	50-200
Diameter of eggs — Dry		mm	1.0-1.5	0.7-1.3	0.4-0.5	1.5-2.0	0.6-0.8	1.5-2.0
Swollen		mm	2.0-2.5	3.7-6.0	0.6-0.7	2.5-3.0	1.0-1.5	3.0-4.0
Number of eggs in 1 kg — Dry		Thousands	700-1000	600-1000	2000	180-220	1500-2200	180-220
Swollen		Thousands	80-120	15-22	600-700	50-80	1000-1300	30-50
Time between fertilisation and hatching (at optimum incubation temperatures)		Daygrade	60-70	24-50	60-70	120-140	110-120	50-60
		Days	3-4	1-2	3	8-15	6-10	$2\frac{1}{2}$-3
Length of newly hatched larvae		mm	4.8-5.0	5.0-5.2	3.5-3.6	8.5-8.7	4.5-5.0	6.4-6.6

Appendix 2. Reproduction Biology of Pond Fish

Specification	Unit		Common carp	Chinese carp	Tench	Pike	Zander	Wels
Propagation temperature	°C		20–24	22–24	22	12–16	12–16	22–24
Water flow rate (per kg/broodstock)	litres/min		1.0–1.5	1.5–2.0	1	1	1–2	1–2
Amount of swollen eggs to be incubated in 7-litre Zuger jar	(a) Number		120–200 000	40–50 000	1 000 000	200 000	0	60 000
	(b) Volume (litres)		1.5–2.5	2–3	1–1.5	1–2	0	1–2
Water flow in Zuger jar	litres/min		0.5–2.0	0.2–0.5	0.5–2.0	0.8–2.0	0	0.8–2.0
Treatment of eggs in Zuger jar	(a) Chemical		Malachite green	Formalin	(as carp)	Malachite green	Malachite green	(as carp)
	(b) Dosage		1:200 000 (5 ppm)	1:10 000		1:40 000	1:40 000	
	(c) Time		5 min	flush every 3–5 hr		flush every 24 hrs	flush every 24 hrs	
Keeping of larvae in 200 litre containers	(a) Time		3–4 days	3–4 days	4–5 days	0	0	0
	(b) Number		400–600 000	400–600 000	500–600 000		(fine mesh cages)	
Water flow in container	litres/min		10–15	10–15	10–15	4–6/cage	4–6/cage	4–6/cage

Appendix 3. Fry Rearing in Farm Ponds

Specification	Cyprinids	Pike	Zander	Wels
Duration of rearing	3–4 weeks	4–6 weeks	4–6 weeks	3–6 weeks
Rearing site:				
(a) Large scale	Pond 100–1000 m²			
(b) Small scale	Different types of tanks or aquaria	(as cyprinids)	(as cyprinids)	(as cyprinids)
First feed in hatchery	(a) Hard-boiled egg suspension (b) Newly hatched *Artemia* nauplii (c) Collected rotifers	Newly hatched *Artemia* nauplii Collected zooplankton	Collected rotifers	Commercially available starter foods Chopped tubifex Collected zooplankton
Size of first feed	100–200 μm	100–200 μm	100–200 μm	100–200 μm
Stocking density in rearing pond	100–600 larvae/m²	5–50/m²	10–30/m²	10–50/m²

	1-100 larvae/litre	10-30/litre	—	50-500/litre
Stocking density in tanks or aquaria	1-100 larvae/litre	10-30/litre	—	50-500/litre
Daily feeding:				
(a) In pond	1 litre of fine mixed meal (30–40% protein) per 100 000 larvae/day	0	0	0
(b) In tanks	*Ad libitum* living organisms	*(Artemia nauplii or collected zooplankton)*		
Percentage survival:				
(a) In pond	30-40	10-50	5-10	30-50
(b) In tanks	50-80	60-80	—	80-90
Size of fry at the end of rearing:				
(a) Length (cm)	2.5-3	3-5	3-4	3-6
(b) Weight (g)	0.2-0.3	—		

Appendix 4. The Yield of Carp (in Monoculture) in Farm Ponds

	Stocking		Survival		Yield
	(fish/ha)	(kg/ha)	(fish/ha)	(%)	(kg/ha)
From larva to one-summer	100 000–200 000	0	10 000–40 000	5–30	200–400
From fry nursed to one-summer	40 000–60 000	8–15	20 000–35 000	50–60	300–700
From one-summer to two-summer	5 000–7 000	100–200	3 000–4 000	50–70	600–800
From two-summer to market-size	600–800	120–200	400–500	50–70	600–700

Appendix 5. The Yield of Carp (in Monoculture) in Farm Ponds: Semi-intensive Management

	Stocking		Survival		Yield
	(fish/ha)	(kg/ha)	(fish/ha)	(%)	(kg/ha)
From larva to fry nursed	1–4 million	0	300 000–2 million	30–60	90–400
From larva to one summer	300 000–600 000	0	25 000–70 000	5–30	400–1 000
From fry nursed to one-summer	60 000–120 000	20–30	35 000–60 000	50–70	900–1 400
From one-summer to two-summer	10 000–15 000	100–300	6 000–10 000	50–70	1 200–1 800
From two-summer to market-size	1 000–2 500	200–500	800–2 000	60–80	1 200–1 600

Appendix 6. Yield of Fish Ponds under Extensive Management (Polyculture)

	Stocking		Survival		Yield
	(fish/ha)	(kg/ha)	(fish/ha)	(%)	(kg/ha)
From larva to one-summer:					
Carp	100 000–150 000	—	10 000–20 000		150–300
Silver carp	50 000–80 000	—	5 000–10 000		50–100
Bighead carp	10 000–20 000	—	1 000–2 000	5–30	20–30
Grass carp	10 000–20 000	—	1 000–2 000		20–30
Total	170 000–270 000	—	17 000–34 000		240–500
From advanced fry to one-summer:					
Carp	35 000–50 000	7–13	16 000–25 000		270–500
Silver carp	10 000–15 000	2–5	5 000–7 000		100–200
Bighead carp	3 000–5 000	0.7–1	1 500–2 500	50–60	25–50
Grass carp	3 000–5 000	0.7–1	1 500–2 500		25–50
Total	50 000–75 000	12–20	24 000–37 000		430–800
From one-summer to two-summer:					
Carp	4 000–6 000	70–150	2 500–3 200		500–600
Silver carp	1 500–2 000	20–50	800–1 400		150–200
Bighead carp	300–500	10–15	150–300	50–70	50–100
Grass carp	300–500	10–15	150–300		50–60
Total	6 100–9 000	110–250	3 600–5 200		750–950
From two-summer to market-size:					
Carp	500–700	100–140	350–500		450–550
Silver carp	200–300	40–60	150–200		150–180
Bighead carp	50–70	10–15	40–50	50–70	50–60
Grass carp	50–60	10–12	30–40		30–40
Total	800–1 000	150–250	570–790		680–830

Appendix 7. The Yield of Fish Ponds under Semi-intensive Management (Polyculture)

	Stocking		Survival		Yield
	(fish/ha)	(kg/ha)	(fish/ha)	(%)	(kg/ha)
From larva to one-summer					
Carp	200 000–500 000	0	20 000–60 000		300–1 000
Silver carp	100 000–300 000	0	10 000–50 000		150–400
Bighead carp	20 000–100 000	0	2 000–10 000	5–30	30–100
Grass carp	20 000–100 000	0	2 000–10 000		30–100
Total	350 000–1 000 000	0	34 000–170 000		510–1 600
From nursed fry to one-summer:					
Carp	60 000–80 000	12–20	30 000–40 000		600–1 200
Silver carp	20 000–30 000	4–10	10 000–18 000		250–500
Bighead carp	5 000–10 000	1–2	3 000–5 000	50–70	60–150
Grass carp	5 000–10 000	1–2	3 000–5 000		40–150
Total	90 000–130 000	18–34	46 000–68 000		1 010–2 150
From one-summer to two-summer:					
Carp	8 000–10 000	100–200	5 000–7 000		1 000–1 300
Silver carp	3 000–5 000	30–100	1 500–4 000		250–500
Bighead carp	500–1 000	10–20	250–600	50–70	80–180
Grass carp	500–1 000	10–20	250–600		80–180
Total	12 000–17 000	150–340	7 000–12 000		1 410–2 160
From two-summer to market size:					
Carp	900–2 000	180–390	700–1 500		1 100–1 400
Silver carp	300–500	60–100	200–400		200–400
Bighead carp	70–150	15–30	40–100	60–80	50–150
Grass carp	70–150	15–30	40–80		50–80
Total	1 350–2 800	270–550	980–2 080		1 400–2 030

Appendix 8. Long-term Treatments in Pond Conditions (without Water Exchange)

Name of disease	Treatment		Fish name and age groups	Remarks
	Name	Dosage		
Gill diseases/necrosis, myxobacteria infections, etc.	(1) Copper sulphate	$0.2–0.3 \, g/m^3$	Older ages of pond fish	The treatments can be repeated over a two-week period.
	(2) Quicklime (CaO or Ca(OH)$_2$)	$10–20 \, g/m^3$		
	(3) Calcium hypochlorite CaCl(OCl).4H$_2$O	$1.0–1.5 \, g/m^3$		This treatment can be used against high ammonia concentrations.
Costia, Chilodinella, Trichodina	Copper oxychloride 3CuO.CuCl$_2$.3H$_2$O	$4 \, g/m^3$	All ages of cyprinids	
White spot	(1) Malachite green	$0.1–0.4 \, g/m^3$	All pond fish at older ages	In cases of high densities of parasites, even these low dosages of malachite green may kill the fish.
	(2) Malachite green	$0.05–0.1 \, g/m^3$ and add fresh water to the pond	Young catfish	
Dactylogyrus, Gyrodactylus, Lernea, Argulus, Piscicola	Dipterex/or other trichlorfons	$0.5–1.0 \, g/m^3$	All age groups of pond fish	

Appendix 9. Rapid Treatments of Fish in Tanks

Name of disease	Treatment			Fish name and age group
	Name	Dosage	Time	
Saprolegnia	Malachite green	1–3 ppm $(1–3 \text{ g/m}^3)$	10–20 min	Cyprinids, catfish at all ages. Younger fish use lower dosages for shorter time.
Costia, Chilodinella, Trichodina	(1) Table salt	2–5% w/v $(20–50 \text{ g/l})$	*5–10 min 30–60 min	Fingerlings, older age groups of all fish.
	(2) Formalin	100–200 ppm	30–60 sec	Advanced fry of cyprinids.
	(3) Table salt	1–1.5% $(10–15 \text{ g/l})$		
Gyrodactylus, Dactylogyrus, other gill and skin flukes	(1) Table salt	2–5% w/v $(20–50 \text{ g/l})$	*5–10 min	All ages and species of pond fishes
	(2) Ammonium hydroxide (NH_4OH) (25%)	500–1000 ppm	0.5–1 min	Ornamental fishes (e.g. goldfish)
	(3) Dipterex	100 ppm	30–60 min	
Learnea, Argulus, Piscicola	(1) Table salt	2–5% w/v $(20–50 \text{ g/l})$	*5–15 min	Older ages of common carp, koi, goldfish, other cyrpinids.
	(2) Dipterex + Potassium permanganate $(KMnO_4)$	100 ppm 20–50 ppm	30–60 min	

* or shorter depending on the condition.

General remarks: Before these treatments, consult with the vet, as the particular environmental factors of the farm or the weak condition of the fish may prevent these treatments from being used. Before large scale treatments are undertaken a test treatment with a few fish is advisable.

Further Reading

Billard, R. (1999) *Carp – Biology and Culture*. Springer, New York.

Hoole, D., Bucke, D., Burgess, P. & Welby, Ian. (2001) *Diseases of Carp and Other Cyprinid Fishes*. Fishing News Books, Oxford.

Horváth, L., Tamas, G. & Tolg, I. (1984) *Special Methods in Pond Fish Husbandry*. Halver Corp., Seattle.

Horváth, L., Tamas, G. & Coche, A.G. (1985) *Common Carp – 1. Mass Production of Eggs and Early Fry*. Food and Agriculture Organisation of the United Nations, Rome.

Horváth, L., Tamas, G. & Coche, A.G. (1985) *Common Carp – 2. Mass Production of Advanced Fry and Fingerlings in Ponds*. Food and Agriculture Organisation of the United Nations, Rome.

Huet, M. (1986) *Textbook of Fish Culture – Breeding and Cultivation of Fish*. Fishing News Books, Farnham.

Jhingran, V.G. & Pullin, R.S.V. (1988) *A Hatchery Manual for the Common, Chinese and Indian Major Carps*. Asian Development Bank, ICLARM, Manila, Philippines.

Pillay, T.V.R. (1990) *Aquaculture – Principles and Practices*. Fishing News Books, Farnham.

Seagrave, C. (ed.) (2001) *The Sparsholt Guide to the Management of Carp Fisheries*. Mitchellwing Publications, Sparsholt.

Index